Next-Gen Manufacturing 2017

Topical Conference at the 2017 AIChE Annual Meeting

Minneapolis, Minnesota, USA
29 October - 3 November 2017

ISBN: 978-1-5108-5784-1

Next-Gen Manufacturing 2017

Topical Conference at the 2017 AIChE Annual Meeting

Minneapolis, Minnesota, USA
29 October - 3 November 2017

AIChE
The Global Home of Chemical Engineers

Printed from e-media with permission by:

Curran Associates, Inc.
57 Morehouse Lane
Red Hook, NY 12571

proceedings.com

Some format issues inherent in the e-media version may also appear in this print version.

Printed by Curran Associates, Inc. (2018)

For permission requests, please contact AIChE
at the address below.

AIChE
120 Wall Street, FL 23
New York, NY 10005-4020

Phone: (800) 242-4363
Fax: (203) 775-5177

www.aiche.org

Additional copies of this publication are available from:

Curran Associates, Inc.
57 Morehouse Lane
Red Hook, NY 12571 USA
Phone: 845-758-0400
Fax: 845-758-2633
Email: curran@proceedings.com
Web: www.proceedings.com

TABLE OF CONTENTS

Some abstracts in this book refer to images that were not made available by the publisher

509411 Predicting Temperature-mediated Solid Form Transformations in Small Molecule Crystals with Molecular Dynamics

Abstract:

Many pharmaceutical compounds can crystallize into more than one solid form, and the overall solubility and bioavailability of a therapeutic compound can vary markedly between different structures. Selecting a stable and soluble solid form is therefore an important step in the drug development process, and an unintended late-stage transformation into a new solid form can severely delay a therapeutic reaching the market. Computational models of a compoundâ€™s solid form landscape can reveal previously unobserved crystal structures and assist in selecting the appropriate form for commercial manufacturing. However, the current approaches for crystal structure predictions (CSP) often neglect the effects of temperature on solid form stability. Such approaches cannot capture temperature-mediated transformations that can occur during product manufacturing and storage.

In this work, we present a method to compute the temperature-dependent stability of small molecule crystals and capture temperature-mediated solid form transformations. All crystal structures of interest are heated from zero Kelvin up to ambient conditions using molecular dynamics simulations. The sets of simulations are combined to construct temperature-dependent free energy profiles for each structure. We demonstrate this technique on multiple polymorphic small molecules of pharmaceutical interest, and show that current point-charge potentials are capable of capturing the large entropy differences between forms which facilitates crystal transformations at high temperature.

In addition, we show that molecular dynamics simulations can yield visual insights into when crystalline systems undergo critical property changes as a result of a change in temperature. These changes include solid-solid transformations, order-disorder transitions, and lattice minima interconversions.

509412 Polymer D-255 Cement Fluid Loss Additive Control Optimization

Abstract:

Incorporated and optimized new designs with D-255 FLAC technology to meet the same performance given by previous additive D-167. The project had the main goal to improve control of water loss from cement to formation by reducing the permeability of the cement filter cake, and increasing viscosity of the aqueous phase. The testing results concluded new technology to have better stability and compressive strength development. The replacement of new technology with a PC/PR of 0.27 will save $81K to Schlumberger with a 50% sales profit increase in five years.

Modeling Paste Transport Systems for FGD Water Disposal

Trent Rogers

Research & Development

SOUTHERN COMPANY

What is Paste?

- Fly ash encapsulated FGD wastewater
- Non-Newtonian, cementitious fluid pumped into landfills
- FGD wastewater disposal technique that is zero liquid discharge (ZLD) compliant

Uncured paste sample vs cured paste sample

Project Overview

- Model the approximate cost of operating a paste system using a rheometer
- Two dominant cost factors:
  1. Energy/capital costs of pumping (head loss)
  2. Raw material cost of paste (mix design)
- No reliable model for rheopectic pseudo-plastics in mining industry

Mining paste backflow system

Rheometer Testing

- Used to characterize non-Newtonian flow
- Two rheometer measurements
  1. Shear rate: angular velocity of spindle
  2. Shear stress: torque of motor
- These measurements can be used to simulate pipe wall interaction

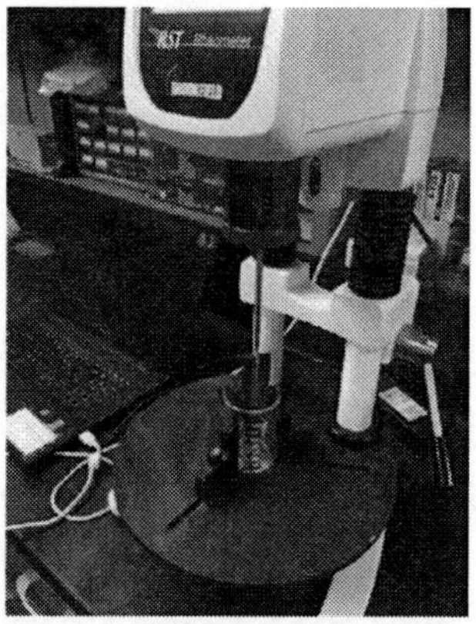

Rheometer Setup

Estimating Head Loss

- Modeling head loss using a rheometer is fairly common; time dependency is often neglected
- Paste is an "apparent rheopectic" fluid due to rapid cementitious reaction
- Modified cement slurry model (Haimoni and Hannent) for rheopectic behavior

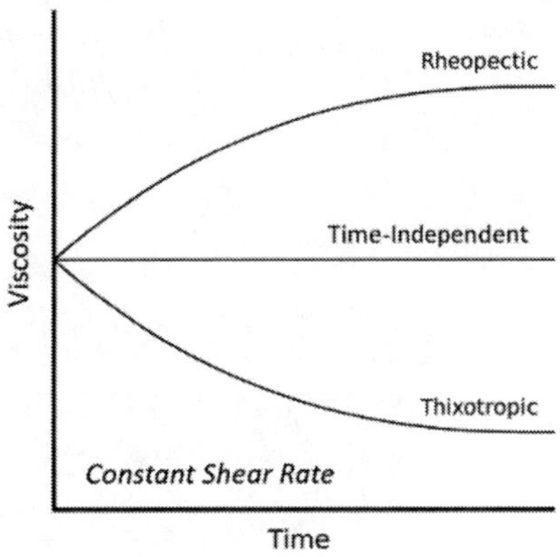

Mix Design Costs

- Chemical additives greatly affect rheology
- Balancing material cost and energy cost is essential
- Mix design can also impact capital costs if a different pump is required

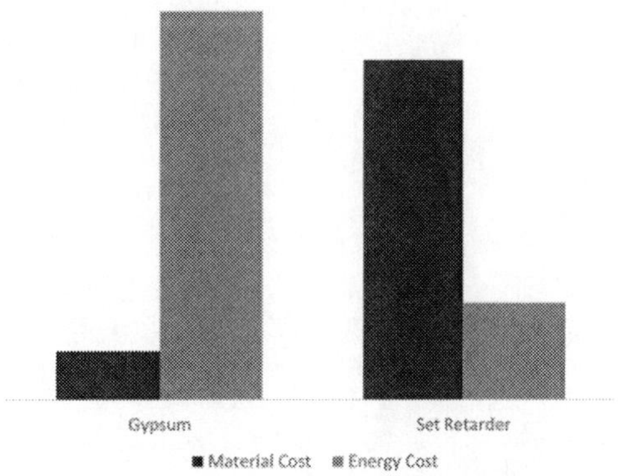

Summary of Modeling Process

1. *Rheometer*
   - Use a designed velocity and rheometer geometry to determine a controlled shear rate
2. *Head Loss Calculation*
   - Use the measured shear stress to simulate the changing pressure loss at the pipe walls
3. *Energy Cost Analysis*
   - Use the pressure loss to determine the energy cost of pumping material at the designed piping distance

Rheometer Results

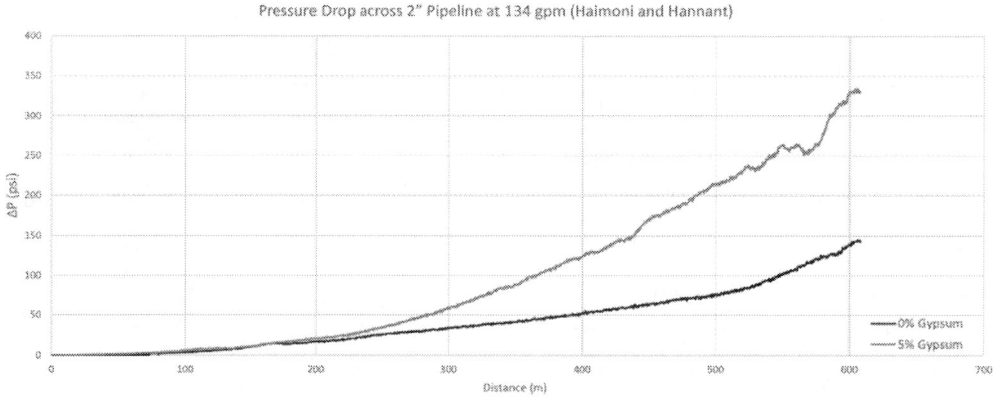

Rheometer Results

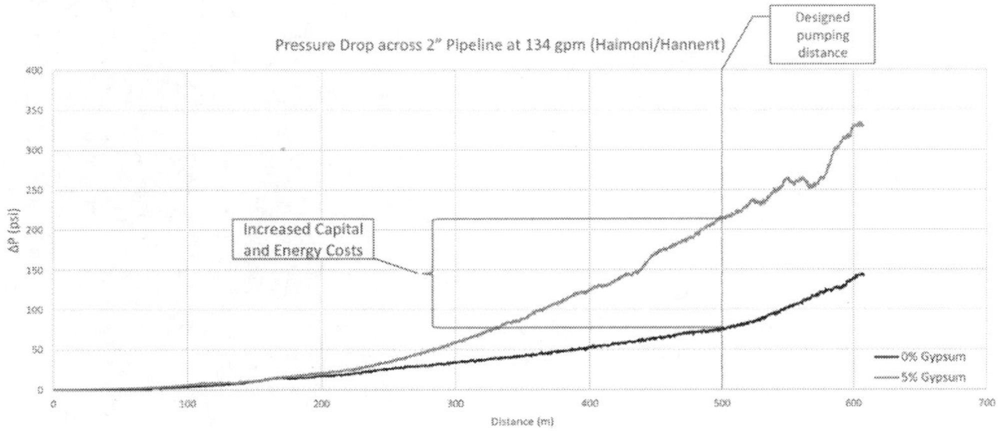

Simulated Case Study: Replacing 5% Gypsum

- Material cost is a constant, but the energy cost increases as time increases
- If pumping beyond the red line, the energy savings outweighs the material savings

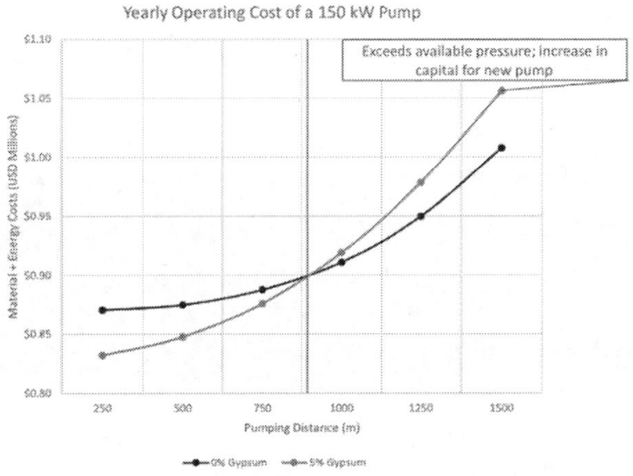

Conclusion

- Next step is confirming these results on a full-scale system
- Simple test that can yield massive amounts of data
- This method can be applied to any variety of time-dependent non-Newtonian fluid

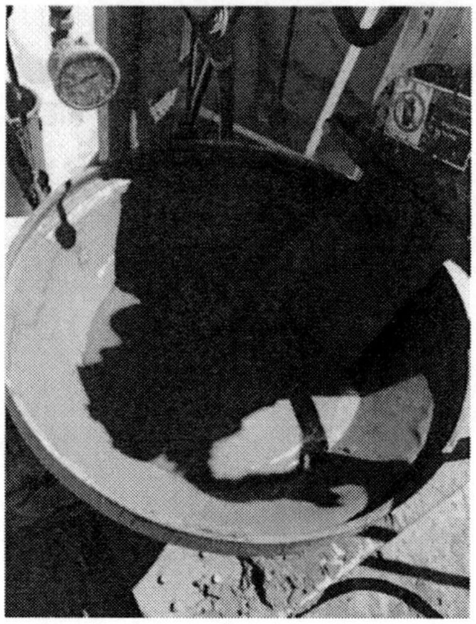

Uncured paste flowing out of mixer

Questions

R&D Research & Development

SOUTHERN COMPANY

511600 Production of Nimesulide Nano-particles via Liquid Anti-solvent Precipitation using Spinning Disc Reactor

Abstract:

This summer I was offered a research fellowship by Indian Academy of Sciences (IAS) to conduct a project on biochemical applications of nanotechnology in Institute of Chemical Technology, Mumbai. According to a study, nearly 40% of the pharmaceutical drugs are poorly water soluble due to which bioavailability of these pharmaceutical drugs is insubstantial and, high dosages of these drugs have to be administered.

My project focused on finding the optimum parameters for producing stable drug nanoparticles in order to increase its dissolution rate. For my study, I chose Nimesulide drug, a non-steroidal anti-inflammatory drug and produced its nanoparticles in a spinning disc reactor using the principles of Liquid-Antisolvent precipitation. The size of the raw drug was found to be 18 microns. The project required a comprehensive knowledge of chemical engineering principles such as transport phenomena, biochemical engineering, reaction engineering and mass transfer. Using these principles, supersaturation of drug was achieved. Spinning disc reactor employed applications of momentum and mass transfer for the production of coated nano-drug. The thickness of boundary layer was also attempted to study.

During my project, I monitored influences of operating parameters such as flow rate of solvent-antisolvent, solvent-antisolvent ratio, nimesulide concentration, disc characteristics and rotating disc speed on the nanoparticles size. At the optimum parameters, particles as small as 190 nm were obtained with a substantial increase in dissolution rate. Samples were analyzed using Dynamic Light Scattering and UV-Vis spectroscopy.

This project is a gateway to finding an efficient method of drug delivery. Using this method, the bioavailability of the drugs could potentially increase which would not only reduce the dosage of the drug administered but also make the drug economically feasible. Production of nano-drug and development of generic pathways for drug delivery could hold the potential cure for epidemic diseases like AIDS, cancer, and tuberculosis.

503000 Autothermal Pyrolysis of Lignocellulose Wastes to Sugars and Other Biobased Products

Abstract:

The goal of this RAPID Institute project, a collaboration between Iowa State University (ISU) and Easy Energy Systems (EES), is to develop a pyrolysis-based Modular Energy Production System (MEPS) for the thermal deconstruction of lignocellulosic biomass into cellulosic sugars and other value-added products. ISU is developing several innovations in process intensification of pyrolysis that dramatically enhances the prospects for distributed processing of biomass to produce cellulosic sugars and other products. The focus of this project during Budget Period 1 include: (1) evaluation by ISU of continuous pretreatment and autothermal pyrolysis of lignocellulosic feedstock in the process intensification of cellulosic sugar production; and (2) design and construction by EES of modules for feedstock handling and autothermal pyrolysis at the 50 ton per day demonstration scale. Successful completion of these two activities will set the stage for design and construction of other modules associated with demonstrating pyrolysis-based MEPS as a viable approach to low-cost cellulosic sugar production.

Autothermal pyrolysis investigated in this project is a specific example of the more general principle of autothermal chemical processing, which balances energy demand from endothermic reactions with exothermic reactions. Autothermal chemical processing has application to other processes for which heat transfer is the bottleneck to process intensification. Although reactions that proceed to chemical equilibrium are relatively easy to operate under autothermal conditions, non-equilibrium processes, like fast pyrolysis, are more challenging to operate autothermally without adversely affecting desired product yields. Our work to date demonstrates that autothermal chemical processing has the potential to simplify reactor design and dramatically increase throughput when heat transfer is the bottleneck for a chemical process.

503004 Dynamic Process Intensification

Abstract:

Process intensification (PI) aims to create smaller, safer and more efficient chemical plants. This objective is often met by combining multiple conventional unit operations in a single physical device. The vast majority of the developments made in the PI field consider primarily the steady state design and operation of such devices. It was noted repeatedly that the dynamics and control of intensified systems are complicated by interactions between the underlying unit operations. In this project, we take the converse perspective of focusing on the dynamics of intensified processes, with the goal of improving productivity. Specifically, we define periodic operating patterns that exploit the interplay between multiple phenomena occurring within intensified devices; on the average, this operating paradigm can lead to improved (time average) performance compared to steady-state operation for the same hardware. We refer to this result as â€œdynamic process intensification.â€ In this talk, formal tools for characterizing the potential for dynamic intensification are briefly described, and a case studyis presented.

503002 Manufacturing Supply Chain Development for Solar Thermochemical Modules

Abstract:

Solar thermochemical modules have been demonstrated with high solar-to-chemical conversion efficiencies for producing syngas via steam methane reforming (SMR). Reaction systems consisting of SMR reactors, heat exchangers, and other potential unit operations are located at the focal point of parabolic solar concentrators to directly convert solar heat to chemical energy. Thus, these inherently modular systems benefit from process intensification technologies, such as microchannel heat exchangers, to reduce system size and weight. Commercialization of this technology is facilitated by reducing equipment costs and establishing manufacturing supply chains, a suitable role for the Modular Manufacturing Focus Area of the RAPID Institute. Process-based cost models for manufacturing components have enabled analyses of alternative manufacturing processes, established the importance of reducing material costs, and incentivized alternative designs. Additive manufacturing methods are being pursued as an alternative for fabricating reactors and heat exchangers. Two initiatives are described for reducing material costs. In the first, efforts are focused on the development of new powder bed fusion techniques capable of producing lower cost refractory alloys compared to Ni-based superalloys. Preliminary results will demonstrate progress toward producing lower cost refractory alloys from less expensive base powder. In the second initiative, high material utilization forming methods are being developed for producing lower-cost high efficiency, compact heat exchangers. The paper will focus on cost modelling methods and tools and how they are being used to direct research efforts toward lowering equipment costs.

503006 RD&D Needs for Sustainable Separations

Abstract:

The road to a sustainable tomorrow is clearer now that innovators from across the chemistry enterprise have begun a collaboration to accelerate the industrial deployment of sustainable alternative separation (AltSep) processes in chemical manufacturing. Rooted in a partnership between the ACS Green Chemistry Institute Chemical Manufacturers Roundtable and the AIChE, AltSep aims to transform the way we apply separation process technology in chemical manufacturing. The first step in this transformation has been the creation of a technology roadmap for advancing the rational design and predictable, widespread industrial application of sustainable AltSep processes. These mass separating agent-based separation processes can slash the energy required for chemical separations through advances in modular process intensification. The roadmap has been built through broad collaboration across industry, academic institutions, and government. This presentation will outline key research, development, and demonstration (RD&D) needs identified by the roadmap.

504195 Sustaining Innovation / Innovating Sustainably

Abstract:

Optimism is at the heart of both innovation and sustainability. Research creates discoveries that lead to inventions, creating innovations that make the world better. R&D starts with the optimistic view that you can create something better. It is an investment in the future, where resources remain for future generations. At Dow, weâ€™ve earned the ability to continue innovating based on our past performance. R&D isnâ€™t left to chance, we take steps to stack the deck by:

- Rigorously manage our R&D portfolio based on quantifiable metrics
- Focus on R&D productivity and timely delivery of technology
- Collaborating to extend access to new inventions and to create customer intimacy
- Application of technical excellence in both product and process design

Portfolio management, built on a well-established foundation, guides our efforts and incorporates sustainability and other metrics to influence decisions. We use the life-cycle impact of our products to guide our efforts. Far too often, hype has surpassed reality, boundary conditions have been neglected, and actions taken that do not truly lead to a sustainable future. Looking across the life-cycle compels decisions to reduce raw material use, energy use and to limit toxics use. Innovation will create a sustainable future.

504196 Creating Chemistry for a Sustainable Future

Abstract:

Imagine a future where we can use our creativity and innovation power to help solve some of the largest global sustainability challenges of our times. Are we doing enough to enable this?

Investing in businesses that seek positive economic, ecological and social impact is becoming imperative. We will be judged by our communities, by our work-force, by the talent pool of Millennials, by investors. We will be judged in the courts of social media and public opinion. We have to act now; authentically, and for the right reasons

Chemical companies are positioned uniquely to solve the global sustainability challenges. But we have to take a holistic look at these challenges; going beyond our typical value chains and including the larger eco-system of stakeholders. That is, going beyond current business models to co-creating solutions.

For example, Chemistry plays an active role in improving energy efficiency of homes. There are many opportunities to co-create solutions to improve this situation. BASF therefore, is partnering with other stakeholders to change state laws and make it easier to update residential building codes. This challenged us to work with new stakeholders: industry partners, architects, city planners, builders, citizens, governments and many more. This is what we mean by co-creation.

BASF is convinced of the importance of the UN Sustainability Development goals to our long-term success. This is not solely about recycling or bio-based raw materials; those are of course important. Moving forward, however, requires us to build upon existing platforms and address critical issues such as transparency, education, diversity and inclusion â€“ to just name a few.

No single company can address all the seventeen UN sustainability development goals on their own. Co-creation is the way forward to address both the overlaps and the gaps.

504198 Continued Delivery on Impactful Sustainable and Innovative Business and R&D Strategies

Abstract:

Expanding our impact through sustainability is a core value at Owens Corning. From the products we make to the way that we make them, sustainability is an essential element of our business. We design innovative solutions that meet expectations of customers/markets and contribute toward addressing significant global issues such as energy conservation, environmental impact and overall quality of life. We consistently leverage our technological expertise to deliver sustainability-driven innovations to the Insulation, Roofing and Composites markets. Our insulation products enhance energy efficiency; cool roof shingles help reduce air-conditioning and power consumption; reinforced lightweight glass-composites deliver superior mechanical properties with lower environmental impact than traditional materials. Our business strategies promote innovative ways to achieve high-performing, greener and healthier products via high-efficiency manufacturing processes with reduced energy and water consumption. The discussion in this panel will highlight Owens Corningâ€™s innovative strides throughout its core businesses, its continued commitment to robust sustainability progress in its handprint and footprint, and the companyâ€™s significant safety improvement.

508958 Smart Manufacturing in the Automobile Industry

Abstract:

General Motors is transforming as customers and the industry move towards new mobility solutions, including electrification, autonomous vehicles and ride and car sharing. These changes drive innovations not only in the vehicles that General Motors makes, but also how it makes them. The companyâ€™s goal is a higher level of data-driven productivity from its global manufacturing network, including active monitoring, analytics and detailed system integration. This presentation will address how GMâ€™s manufacturing process is evolving in the digital era.

508959 Smart Manufacturing in Chemical Industries

Abstract:

Smart manufacturing provides a revolutionary ways for the next generation of production in the chemical manufacturing industries. One of the unique advantages of smart manufacturing is the creation of a knowledge base to reduce the operating and capital costs of chemical processes. The Smart Manufacturing Leadership Coalition (SMLC) has collectively developed and is pursuing a comprehensive technology and business roadmap to develop and implement Smart Manufacturing capabilities that will enable next-generation economic, energy, sustainability and EH&S manufacturing performance and competitiveness. (SMLC, 2011). We viewed Corningâ€™s concept of smart manufacturing as a combination of multiple capabilities (integrating measurements, controls, computing platforms, modeling and analytics) to improve process/product development, manufacturing performance, and reduce manufacturing cost within a plant and across the supply chain as very encompassing. (Michelle Pastel, Corning 2016). The challenges and management considerations in smart manufacturing will be presented.

508960 Innovations in Chemical Engineering: Automation of the Factories of the Future and the Impact of Internet of Things (IoT) from the Control of the machinery in factories to Home Appliances

Abstract:

The industrial world is primed for digitization and General Electric is one of he companies leading the application of Internet of Things to the manufacturing environment. Advanced software is available that provides access to secure industrial strength cloud environment for different machine data to be oaded in a secure environment for analytics to be applied for efficiency in factories of the future. The industrial Internet (HOT) will also apply artificial intelligence, virtual reality, and augumented reality in new manufacturing as well as other uses like the control of home appliances. This paper will focus on the opportunities for chemical engineers to lead these innovations from process control to novel IoT sensors development, and the need for advanced materials to enable transformational changes in the Internet of Things platform.

508961 Reversing the Tide in Science, Engineering, Technology, and Science (STEM): Academically Gifted African American Students in Historically Black Colleges and Universities (HBCU)

Abstract:

What started off as a national concern has escalated to the point of crisis (Bonner, Alfred, Lewis, Nave & Frizell, 2009; Chubin & Malcolm, 2008; Hrabowski, Summers & Hrabowski, 2006). The current state of affairs regarding African American participation in STEM disciplines has become one of the thorniest issues of contemporary focus. This presentation highlights data collected for a NSF Education Research grant. The focus of this HBCU-UP project was to uncover factors contributing to the success of academically gifted (high-achieving) African American students enrolled in engineering programs at the 12 ABET accredited four-year institutions in the U.S. Specifically, this research highlights qualitative case study data including key categories and themes identified as impacting the academic success of this high-achieving cohort across each of the 12 institutions included in this study. Based on the findings of this study, recommendations are provided to increase the success of African American gifted students in STEM fields at HBCUs.

508962 Blacks in Science, Engineering and Medicine: Struggles that Continue, Struggles that are Growing, and Possible Solutions

Abstract:

Blacks in many ways face a crisis in regards to numbers of individuals pursuing medicine, engineering and science. For Instance numbers of Black men matriculating into medical school are now lower than they were almost 40 years ago. The causes are manifold. Solutions are demanded.
To quote Baldwin 'Not everything that is faced can be changed, but nothing can be changed until it is faced.'

The American Institute of Chemical Engineers and its Minority Affairs Committee should take a leadership role in advancing solutions to address what is a growing crisis. Next steps are outlined.

RAPID

Transforming Process Industries

Breakthrough Innovation to Transform the Process Industries

Karen Fletcher
CEO - RAPID

Jim Bielenberg
CTO - RAPID

October 30, 2017

ManufacturingUSA

Manufacturing USA Overview

Vision

U.S. Global leadership in Advanced Manufacturing

Mission

Connecting people, ideas, and technology to solve industry-relevant advanced manufacturing challenges, thereby enhancing industrial competitiveness and economic growth, and strengthening our national security.

Goals

1. Increase competitiveness
2. Facilitate technology transition
3. Accelerate the manufacturing workforce
4. Ensure stable and sustainable infrastructure

www.ManufacturingUSA.com

2

RAPID
Transforming Process Industries

Manufacturing USA

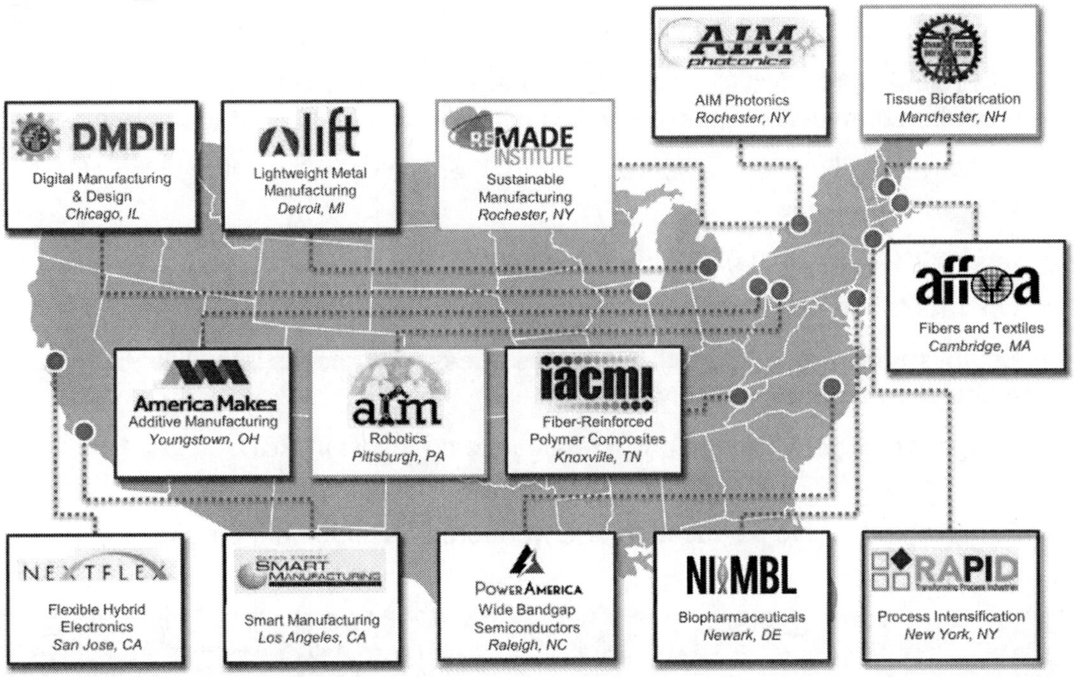

3

RAPID
Transforming Process Industries

RAPID's Industry-Led Vision

A dynamic network of partners who collectively build a sustainable ***ecosystem*** *that:*

... researches, develops and advances new technologies for modular chemical process intensification

... builds a strong portfolio of R&D projects & educational tools

... delivers dramatic reductions in energy, environmental footprint, capital and operating cost

... makes U.S. Manufacturing and our workforce more competitive

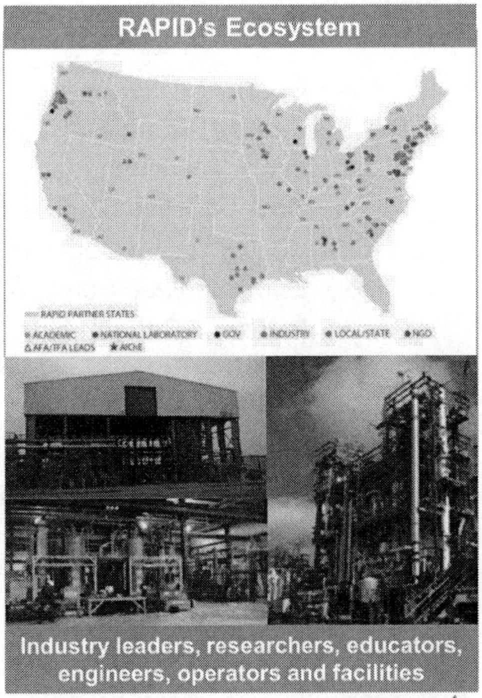

4

RAPID
Transforming Process Industries

Our Mandate

- Lead a national effort to research, develop and demonstrate high-impact modular chemical process intensification solutions for U.S. Manufacturing.
- Actively build RAPID membership through an inclusive and attractive value proposition.
- Leverage $70 million of federal funding with cost share from members.
- Operate the Institute to benefit a wide range of stakeholders.
- Establish an infrastructure that enables access to process intensification resources, tools, expertise, and facilities.
- Bring together private and public entities to co-invest in R&D, commercialization, and deployment of innovative technologies.
- Establish a technical education and workforce development program.

"The goal for these Institutes is to revitalize American manufacturing and support domestic manufacturing competitiveness."
— U.S. DOE

RAPID
Transforming Process Industries

DOE Performance Metrics

- Energy Efficiency
- Energy Productivity improvement
- Intensification in Individual Process Modules
- Cost-Effective Manufacturing of Modules
- Cost Effective Deployment
- Enabling R&D Portfolio
- Industrial Partnerships
- Pathway to Self-Sustainment
- Train the Trainers
- Educate Students
- Annual Planning Process
- Industrial Roadmap
- Emerging Supply Chain
- Diversity of Firms and Individuals in the Eco-System

6

RAPID
Transforming Process Industries

52 Members and growing

Compact Membrane Systems · TEXAS TECH UNIVERSITY · PSE · Chemstations · IOWA STATE · United Technologies Research Center · DOW · ExxonMobil · Power+Energy · easy energy · Georgia Institute of Technology · Swift · Savannah River National Laboratory · SR Southern Research · TEXAS The University of Texas at Austin · gti Gas Technology Institute · TEXAS A&M · AUBURN UNIVERSITY · DUPONT · PRAXAIR · EcoCatalytic Technologies · Lubrizol · American Chemistry Council · aspentech · University of Delaware · Pacific Northwest National Laboratory · Agenda 2020 · Oregon State University OSU · Idaho National Laboratory · The University of Arizona · USC · ACS Green Chemistry Institute Pharmaceutical Roundtable · ACS Green Chemistry Institute Chemical Manufacturers Roundtable · BgtL LLC · IntraMicron · VPE · University of Louisville · University of Minnesota · Case Western Reserve University · West Virginia University · FLUOR · University of Kentucky · Kore Infrastructure · NREL · University of South Carolina · Metcer

RAPID
Transforming Process Industries

Technical Focus Areas

Chemical Commodity Processes

- Develop guidelines for integration of novel reaction and separation modules
- Validate design tools for process intensification

Renewable Bio Products

- Prototype and scale novel bio-conversion processes
- Improve energy & capital efficiency of existing and emerging processes

Natural Gas Upgrading

- Maximize impact through transfer of technologies and learnings across industries
- Mature and demonstrate transformational and enabling technologies for gas utilization

8

RAPID Transforming Process Industries

Technical Focus Areas

Modeling and Simulation

- Develop methods & tools for design, optimization, and intensification across multiple length and time scales
- Establish methods for intensification of dynamic/periodic operations

Intensified Process Fundamentals

- Advance inherently energy efficient separation processes & reaction platforms
- Develop fundamentals for multifunctional modules such as hybrid separation/ reaction schemes

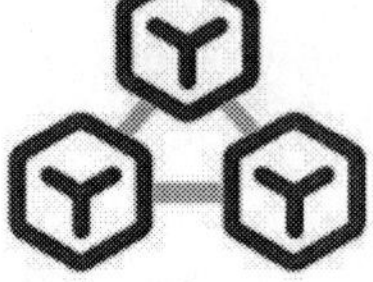

Module Manufacturing

- Standardize modules and components to drive demand and capital investment within the supply chain
- Lower the cost of PI equipment using advanced manufacturing technology

9

RAPID Transforming Process Industries

2017 Accomplishments

- Completed DOE negotiations & all required documentation
- Stood up the Institute, with strong support/services from AIChE
- Established membership base
- Identified, negotiated and launched 4 Jump Start projects
- Formed Education & WFD Committee and completed a roadmap
- Completed focus area roadmaps with 100+ participants
- Executed call for pre-proposals and call for proposals
- Designed and launched a new website www.aiche.org/rapid
- Hired CTO (Mar), Technology Manager (Oct), Comms Leader (Nov)

10

RAPID
Transforming Process Industries

A Public-Private Partnership Between U.S. DOE EERE & AIChE®

Technology Gaps

RAPID Roadmap

Chemical & Commodity Processing

- **Rxn/Separation**
  Novel reaction/separation schemes that are scalable and drive process efficiencies (e.g., membrane or sorption enhanced reactors). Applications include processing light paraffins to olefins, increase propylene yield vs conventional processes, hydrogen production, managing oxygen supply to reactions, etc.
- **Non-Thermal Drivers**
  Use of alternative, non-thermal driving forces to activate chemical systems at the appropriate (atomic/molecular) scales.
- **Batch Systems**
  Intensification schemes for batch systems. Transferring concepts largely developed for continuous processes to the batch realm could result in increased productivity/lower cost for specialty/fine chemicals.

Natural Gas Upgrading

- **Selective Conversion**
  Concepts that show dramatic increases in desired product yields via fundamental improvements in catalysis, heat and mass transfer, and process concepts. This could include include alternative energy inputs and/or the use of novel reaction systems.
- **Separations**
  Energy efficient separations technology to purify the reaction product mix, condition the feed in preparation for conversion, and to generate co-reactants for participation in natural gas conversion gas.
- **Process Consolidation**
  Process consolidation and modularity to reduce total installed cost by reducing the total number of unit operations and by reducing the amount of field fabrication.

Renewable Bioproducts

- **Primary Separation**
  Technologies to reduce energy demand in primary separation process steps designed to recover organic molecules and biomass components from water.
- **Water management**
  Low capital and energy intensive solutions for dewatering and drying of biomass feedstocks, water removal and drying in pulp and paper process, and drying and removal of low levels of residual water from end products.
- **Couple Rxn/Spn/Hxt**
  Use of novel chemistries and MCPI strategies to couple heat transfer and reaction in thermal processing of biomass and/or novel applications of reactive separation technologies in biological conversion technologies such as fermentation.

Intensified Process Fundamentals

- **Scale-Out Methods**
  Scale out methodologies and models to predict performance of alternative energy input approaches for reactions and mixing and determine the suitable scale for modular manufacturing.
- **Fundamental Data Acquisition/Modeling**
  Approaches to address key issues with lack of data on fluxes, adsorption, and catalyst kinetics for wide classes of materials, enabling model development and experimental testing of novel materials as adsorbents, membranes, catalysts and their integration.
- **Predictive Models**
  Modeling capabilities to screen concepts and configurations of all types and predict optimal structures.

Modeling & Simulation

- **PI Software Tools**
  Software tools for integrated reaction and/or separation processes and/or cyclic process such as pressure swing adsorption (PSA) or temperature swing adsorption (TSA). Such tools must be widely accessible and capable of integrating MCPI solutions with existing unit operations.
- **Data Availability**
  Modeling approaches coupled with data generation and/or analysis to generate databases of physical parameters enabling design with mass separating agents.
- **PI Assessment Tools**
  Tools to assess safety, sustainability, and control in PI and MCPI applications, including tools that address unique issues of uncertainty and reduced control variable options that are present in PI and MCPI applications.

Module Manufacturing

- **Intensified Components**
  Intensified components that drive down the cost of module pre-assembly, transportation, and installation, while driving significant energy savings in chemical processes.
- **Standard Designs**
  Design approaches that limit the amount of non-recurring engineering — during systems integration and installation — needed to support customized modules. This could include standard modules that enable economies of mass production and/or designs that enable incremental capacity additions.
- **Distributed Processing**
  Module design and manufacturing approaches to enable distributed chemical processing. These will provide new paths to capital cost reduction and innovative techniques for maintenance and remote access and monitoring.

www.aiche.org/rapid

RAPID
Transforming Process Industries

A Public-Private Partnership Between U.S. DOE EERE & AIChE®

RAPID Education & Workforce Development Roadmap

OBJECTIVE: Establish a technical education and workforce development program that will leverage existing resources to train and educate the workforce, who can then research, develop, design and operate processes that incorporate new process intensification (PI) technologies and modular process designs widely within U.S. industry.

Undergraduate Students

GOAL
Create PI & MCPI awareness in future generations of engineers by augmenting the standard (ABET accredited) engineering curriculum with modules and educational content introducing PI & MCPI topics relevant to industry.

WHAT SUCCESS LOOKS LIKE
500 students* trained in PI & MCPI per year by 2019.

METHOD
eLearning course modules, face-to-face courses in conjunction with major meeting, Intern program.

Graduate Students

GOAL
Generate PI & MCPI expertise so that graduate students can connect the relevance of PI & MCPI research to industrial applications and create new opportunities and applications for MCPI.

WHAT SUCCESS LOOKS LIKE
500 students* trained in PI & MCPI per year by 2019.

METHOD
eLearning course modules and face-to-face courses in conjunction with major meeting.

Faculty

GOAL
Update and educate university faculty on PI & MCPI concepts and introduce eLearning modules created by RAPID for integration into their curriculum.

WHAT SUCCESS LOOKS LIKE
50 education/training professionals per year in MCPI by 2019.

METHOD
Face-to-face course in conjunction with major meeting on the Fundamentals of PI & MCPI.

Technicians/Operators

GOAL
Introduce technicians and operators to PI & MCPI, provide awareness of new processes that include PI & MCPI as well as emphasizing process safety.

WHAT SUCCESS LOOKS LIKE
At least 50 professionals** per year in MCPI by 2019.

METHOD
eLearning course modules.

Professional Engineers R&D | Design | Production

GOAL:
Train the current generation of engineers on how to utilize and deploy PI & MCPI technology in order to improve process efficiency.

WHAT SUCCESS LOOKS LIKE:
At least 50 professionals** per year in MCPI by 2019.

METHOD
eLearning course modules, face-to-face courses in conjunction with major meeting, and promotion through Manufacturing Extension Partnerships (MEPs).

Management

GOAL
Create high level awareness of the benefits of PI & MCPI from a techno-economic analysis perspective.

METHOD
Presentation material (includes relevant case studies, sustainability, and safety), short videos aimed at executives.

PI & MCPI Body Knowledge Drives Content Development Topics and Priority

Crosscutting areas identify content topics, levels of understanding for each audience member as well as high-priority topic areas. It includes four sub-sections:

- PI & MCPI Overview
- Modeling & Simulation
- Intensified Process Fundamentals
- Module Manufacturing

Chemical & Commodity Processing · Natural Gas Upgrading · Renewable Bioproducts

Application areas drive relevancy for content and provide industrial applications, examples and case studies within content and include the three sub-sections above.

* Students defined as undergraduate and graduate students.
** Professionals defined as technician, operator or engineer.

www.aiche.org/rapid

RAPID Transforming Process Industries

2017 Call for Proposals

- Overall Results
  - 73 submissions
  - 42 "lead" institutions
  - ~$30M (fed) request for 2018
- Proposals review process completed during Sept/Oct
- Announcement of selected projects November 2nd, 2017

# of Submissions	Submissions*
Chemical Commodity Processes	13
Intensified Process Fundamentals	13
Modeling and Simulation	6
Module Manufacturing	7
Natural Gas Upgrading	18
Renewable Bioproducts	16
	73

Summary Data - RAPID 2017 Proposal Call

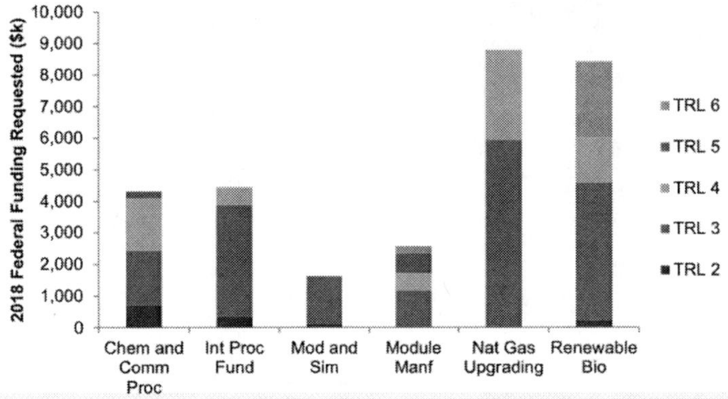

13

RAPID Transforming Process Industries

Dynamic Intensification of the Operation of Dividing Wall Columns

DOW

TEXAS The University of Texas at Austin

Innovation:

- Demonstrate that a dynamically controlled process (non-steady state) creates opportunity for improved efficiency

Project Outcomes

- Develop theoretical basis for periodic operation of intensified processes
- Experimental proof of concept using dividing wall column (DWC) pilot plant available at UT

Institute Target Addressed

- Develop tools and technologies to reduce the cost of deploying modular chemical process intensification in existing processes

Fall 2017 Update

- Theoretical framework for general value/screening of periodic operation tested
- Initial screening of DWC periodic operation assessed

14

Adsorptive Nitrogen Rejection from Natural Gas

RAPID Transforming Process Industries

PRAXAIR

Georgia Institute of Technology

Innovation

- Utilize breakthrough rate dependent adsorbent to modularize conventional PSA process

Project Outcomes

- Field demonstration of portable, modular PSA system

Institute Target Addressed

- 1,000 hours of pilot scale operating time for a modular and intensified process with reduced capacity , improved energy efficiency, and lower emissions/environmental waste

Fall 2017 Update

- System installation, shakedown, and testing to demonstrate adsorbent performance in a PSA pilot system underway
- Manufacturability studies progressing - potential adsorbent manufacturer identified.
- Early 2018 tasks - additional work on TEA and H2S tolerance testing at Georgia Tech

15

Autothermal pyrolysis of lignocellulose wastes to sugars & other biobased products

RAPID Transforming Process Industries

easy energy SYSTEMS

IOWA STATE

Innovation

- Couple intensified reaction + heat transfer design and novel process chemistry to modularize sugar production from biomass

Project Outcomes

- Establish MCPI benefits of autothermal pyrolysis, leading to design of a commercial scale modular system

Institute Target Addressed

- MCPI RD&D resulting in a doubling of energy productivity ($ Returned/Energy Consumed) by a combination OPEX and CAPEX improvements.

Fall 2017 Update

- Demonstrate sugar yields in excess of BP1 targets on 1 kg/h pyrolysis reactor under auto thermal conditions
- Completed continuous pilot scale pretreatment of multiple feedstocks
- Continuing to work on longer time pilot scale testing

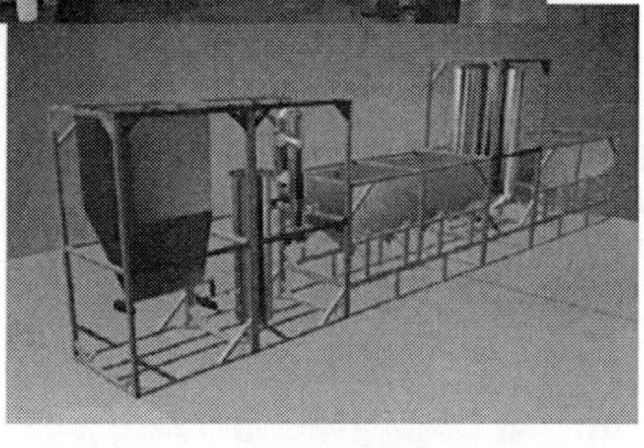

16

Mfg Supply Chain Development for the Modular Solar-Thermochemical Conversion Platform

RAPID Transforming Process Industries

Oregon State University OSU

Pacific Northwest NATIONAL LABORATORY

Innovation

- Coupling material processing approaches and manufacturing techniques that are uniquely well suited to small scales to generate low cost modular components

Project Outcomes

- Develop novel additive manufacturing techniques and materials that enable modular designs

Institute Target Addressed

- Scale-out manufacturing MCPI, with a unit cost reduction of >20% with each doubling in cumulative module manufacturing production

Fall 2017 Update

- OUS has developed unvalidated costs models for key manufacturing steps
- PNNL has established cost targets for key components
- OSU is working in-house capability to test new designs based on selective alloying
- Targeting 1Q2018 for new tool completion in support of novel manufacturing techniques

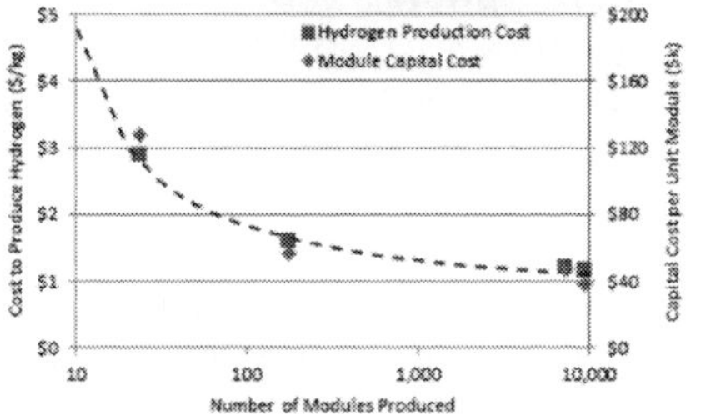

Estimated capital and product costs for individual STARS modules as a function of module mass production.

Plans for 2018

RAPID Transforming Process Industries

- Build the RAPID community
  - Member profiles, webinars, newsletters, website, networking opportunities
  - Conferences / face-to-face meetings (AIChE Spring / Annual)
- $30+M invested in R&D projects (fed funds + cost share)
- DOE (external) peer review
- Updated roadmaps; Gaps and new areas explored
- Project calls
- Training courses and educational webinars
- Increase membership and "reach"
  - New addition to our team: State MEPs*

* MEP = Manufacturing Extension Partnership

RAPID
Transforming Process Industries

To learn more about RAPID:

www.AICHE.org/RAPID

KareF@aiche.org

JameB@aiche.org

19

493251 Modeling and Simulation â€“ a Key Component in Enabling Process Intensification

Abstract:

The RAPID Manufacturing Institute aims to enable the development, de-risking, and implementation of process intensification across the chemical process industry. We will discuss critical needs in modeling and simulation in this area by highlighting the successes and challenges associated with using modeling and simulation to drive RAPIDâ€™s aims.

493349 Modeling and Simulation Challenges for Process Intensification

Abstract:

Modeling and Simulation Challenges for Process Intensification

Efstratios N. Pistikopoulos[1,2], David Sholl[3], M. M. Faruque Hasan[1,2], Salih E. Demirel[1,2] and Yuhe Tian[1,2], (1)Artie McFerrin Department of Chemical Engineering, Texas A&M University, College Station, TX, (2)Texas A&M Energy Institute, Texas A&M University, College Station, TX, (3)School of Chemical & Biomolecular Engineering, Georgia Institute of Technology, Atlanta, GA

The recently established Rapid Advancement in Process Intensification Deployment (RAPID) Manufacturing Institute aims to transform the U.S. industry by developing the next generation of modular, intensified, and innovative manufacturing technologies and processes. The RAPID Modeling and Simulation Focus Area will in particular address the development of strategies and tools for the multi-scale representation, modeling, optimization, and control of intensified systems. In this presentation, we provide an overview of the challenges towards a systematic synthesis, modeling, and simulation approach for process intensification, as well as an exclusive review of process intensification synthesis approach methods in the open literature, such as the ones included in [1-5]. We also introduce an integrated framework for process intensification incorporating operability, control and safety (illustrated in Figure 1).

Figure 1: An Integrated Framework for Process Intensification Incorporating Operability, Control and Safety

[1] Demirel, S. E., Li, J., & Hasan, M. F. (2017). Systematic process intensification using building blocks. *Computers & Chemical Engineering*. http://doi.org/10.1016/j.compchemeng.2017.01.044.

[2] Baldea, M. (2015). From process integration to process intensification. *Computers & Chemical Engineering*, *81*, 104-114.

[3] Tula, A. K., Babi, D. K., Bottlaender, J., Eden, M. R., & Gani, R. (2017). A computer-aided software-tool for sustainable process synthesis-intensification. *Computers & Chemical Engineering*. http://doi.org/10.1016/j.compchemeng.2017.01.001

[4] Papalexandri, K. P., & Pistikopoulos, E. N. (1996). Generalized modular representation framework for process synthesis. *AIChE Journal*, *42*(4), 1010-1032.

[5] Ismail, S. R., Proios, P., & Pistikopoulos, E. N. (2001). Modular synthesis framework for combined separation/reaction systems. *AIChE Journal*, *47*(3), 629-649.

494520 The Sustainable Synthesis-Design-Intensification of Chemical and Biochemical Processes

Abstract:

With the demands for efficient utilization of energy and water, better protection-control of the environment and correct-consistent use of raw materials, it is no longer sufficient to convert a selected raw material to a desired product, making profit and ensuring operability. The synthesized-designed process must now also be sustainable, incorporating not only economic feasibility but also environmental impact constraints, energy-water usage and/or waste constraints, resources availability constraints, product demand-price constraints and, many more. That is, the demands for new, innovative and sustainable processes need to be addressed. Process intensification has been identified as an important means to achieve this demand through integration with process synthesis-design in the early stage development of any chemical and biochemical process.

The presentation will highlight the issues related to defining the measures of sustainability through which new alternatives can be evaluated. Here, emphasis is given to the concept of sustainability, which is relative and therefore, a reference case first needs to be established. That is, compared to the base case, how sustainable are the new and innovative chemical and biochemical process designs?

A systematic multi-stage and multi-dimension computer aided framework which provides the necessary work-flow, the corresponding data-flow, the associated methods and the needed computational tools will be presented.

Through case studies involving chemical and biochemical processes, the application of the framework and the associated methods-tools to determine more sustainable solutions will be illustrated. It will be shown that significant improvements and non-tradeoff sustainable can be obtained through this integrated approach of combining process intensification as an essential element of process synthesis-design.

2017 AIChE Annual Meeting

October 29 - November 3, 2017
Minneapolis Convention Center, Minneapolis, MN

Democratizing Energy Technology

Dane A. Boysen, PhD
Modular Chemical, Inc.

October 30, 2017

Overview

- How did RAPID get established?
- Why do we go big?
- Why is scale-up a problem?
- How do we break the hegemony of scale-up?

2011-2014

arpa·e
Advanced Research Projects Agency • ENERGY

2015-2016

gti®
GAS
TECHNOLOGY
INSTITUTE

Innovation Challenge

FINANCING ECONOMIES OF SCALE

GTL PLANT

(boe/d)	($/bpd)	plant cost
1	47,000k	$47M
10	10,000k	$100M
100	2,200k	$220M
1k	470k	$470M
10k	100k	$1,000M

AMMONIA PLANT

(ton/d)	($/tpd)	plant cost
1	10,000k	$10M
10	4,500	$25M
100	2,000k	$200M
1k	900k	$900M
10k	400k	$4,000M

calculated from "2/3 scaling law"

Innovation Challenge

FINANCING ECONOMIES OF SCALE

GTL PLANT

(boe/d)	($/bpd)	plant cost
1	47,000k	$47M
10	10,000k	$100M
100	2,200k	$220M
1k	470k	$470M
10k	100k	$1,000M

AMMONIA PLANT

(ton/d)	($/tpd)	plant cost	
1	10,000k	$10M	R&D demo
10	4,500	$25M	
100	2,000k	$200M	pilot demo
1k	900k	$900M	commercial
10k	400k	$4,000M	

calculated from "2/3 scaling law"

R&D Spending

OFFICE OF FOSSIL ENERGY BUDGET

U.S. DEPARTMENT OF ENERGY

FY 2015	$561M
FY 2014	$570M
FY 2013*	$495M
FY 2012*	$337M
FY 2011	$434M
FY 2010	$660M
FY 2009	$876M
FY 2008	$465M

Pilot chemical and power plants require **more than $200M**

How can we innovate with vastly inadequate funding?

*continuing resolution

FY2016 U.S. Senate Energy and Water Appropriations Bill

"The Committee recognizes that stranded, flared, and vented natural gas is the result of low natural gas prices that make transporting it uneconomic. As topics for additional Clean Energy Manufacturing Institutes are evaluated, the Secretary is encouraged to consider modular chemical processing as a way to address the issue of natural gas flaring and enable advanced manufacturing applications in the oil and gas industry."

https://www.congress.gov/congressional-report/114th-congress/senate-report/54/1

RAPID

U.S. DEPARTMENT OF ENERGY | Energy Efficiency & Renewable Energy

ADVANCED MANUFACTURING OFFICE

RAPID ADVANCEMENT IN PROCESS INTENSIFICATION DEPLOYMENT INSTITUTE

December 2016, the US Department of Energy announced the establishment of the 10th Manufacturing USA Institute, representing a critical step in the federal government's effort to double U.S. energy productivity by 2030. The Rapid Advancement in Process Intensification Deployment (RAPID) Institute is focused on addressing the barriers listed above to enable the development of breakthrough technologies to boost energy productivity and energy efficiency through manufacturing processes in industries such oil and gas, pulp and paper and various domestic chemical manufacturers. RAPID will leverage approaches to modular chemical process intensification (MCPI)—such as combining multiple process steps such as mixing, reaction, and separation into single more complex and intensified processes—with the **goal of improving productivity and efficiency, cutting operating costs, and reducing waste.**

https://www.aiche.org/rapid/about

RAPID Successes

1. Quick to get up and operational
2. Quality R&D projects in the que
3. Good industry engagement

Challenges for RAPID

1. No clear problem statement (goal ≠ problem)
2. No plan for establishing standards (modular = standards)
3. Not clear how progress will be tracked (need baseline)

RAPID Challenges

1. No clear problem statement (goal ≠ problem)
2. No plan for establishing standards (modular = standards)
3. Not clear how progress will be tracked (need baseline)

Why do we go big?

Energy Technology Today
PRICE TAG
$1-10 BN
PRICE TAG
$0.5-5 BN
PRICE TAG
$5-50 BN
COAL POWER PLANT
AMMONIA PLANT
GTL PLANT

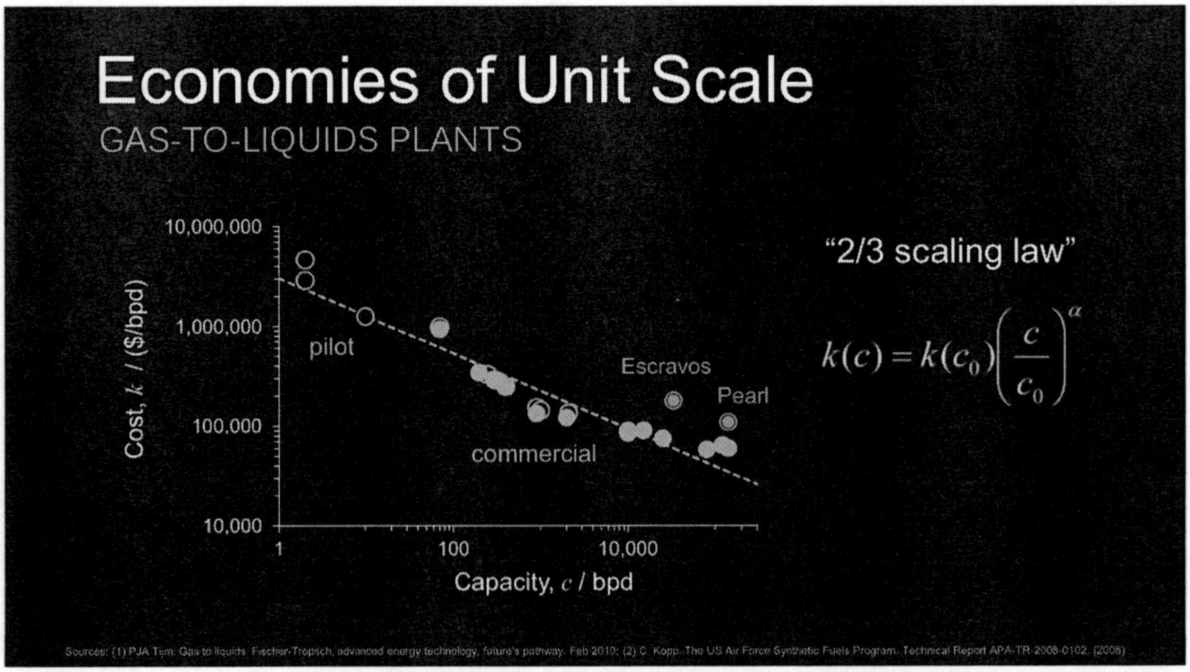
Economies of Unit Scale
GAS-TO-LIQUIDS PLANTS
10,000,000
1,000,000
100,000
10,000
Cost, k / ($/bpd)
pilot
commercial
Escravos
Pearl
1
100
10,000
Capacity, c / bpd
"2/3 scaling law"
$k(c) = k(c_0)\left(\frac{c}{c_0}\right)^{\alpha}$
Sources: (1) PJA Tijm. Gas to liquids. Fischer-Tropsch, advanced energy technology, future's pathway. Feb 2010; (2) C. Kopp. The US Air Force Synthetic Fuels Program. Technical Report APA-TR-2008-0102. (2008)

Why do we scale-up?

ARGUMENT

capital cost $\propto$ area [L^2]
capacity $\propto$ volume [L^3]
capital cost / capacity $\propto$ [L^2]/[L^3]

FLAW

pipe thickness ignored
hoop stress (constant) $\propto t / r$
capital cost $\propto l \cdot r^2$
capital cost / capacity $\propto$ constant

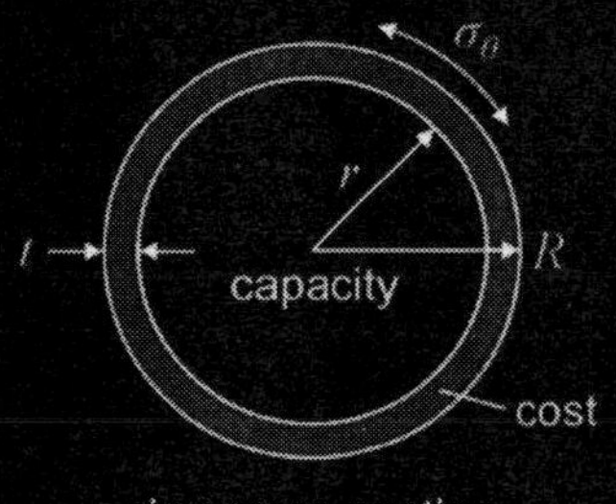

pipe cross-section

Why is scale-up a problem?

#1

BARRIER TO INNOVATION

#2

HUGE CAPITAL NEEDS

MEGA-PROJECT ECONOMICS

RAND Study:

- 52 mega-projects
- $0.5B and $10B (1984 dollars)
- average over budget = 90%

Source: E.W. Merrow, Understanding the outcomes of megaprojects: a quantitative analysis of very large civilian projects, The RAND Corporation, Santa Monica, CA, 1988.

Capital Resources

U.S. COMPANIES BY REVENUE, 2007

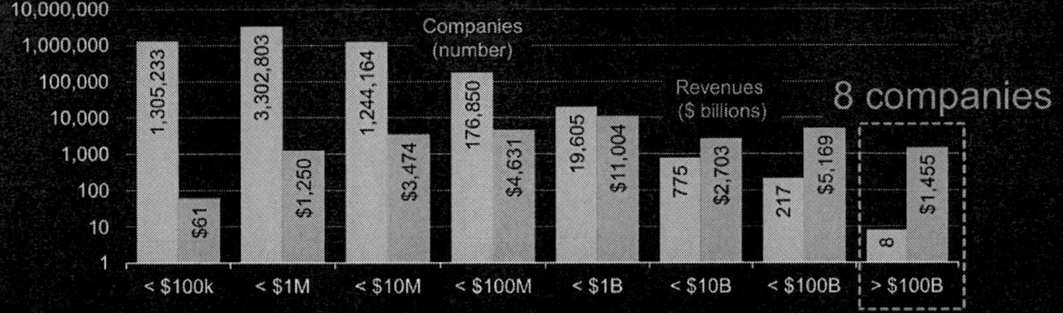

How many U.S. companies can finance a $1 billion plant?

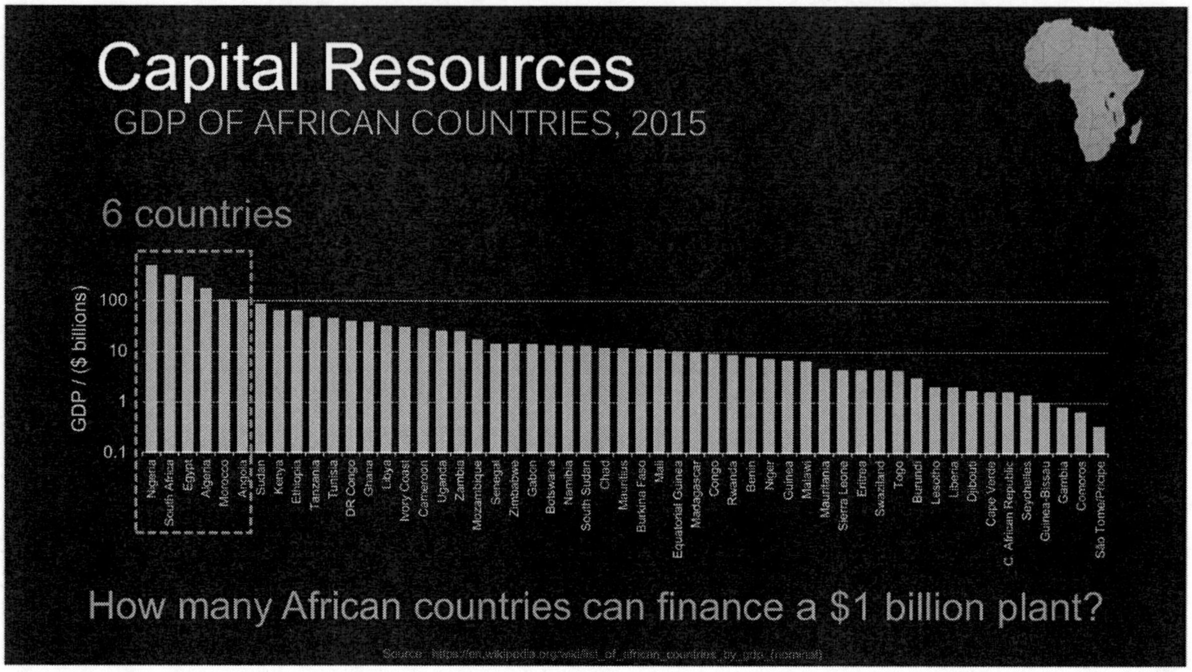

Capital Resources
GDP OF AFRICAN COUNTRIES, 2015
6 countries
GDP / ($ billions)
100
10
1
0.1
Nigeria
South Africa
Egypt
Algeria
Morocco
Angola
Sudan
Kenya
Ethiopia
Tanzania
Tunisia
DR Congo
Ghana
Libya
Ivory Coast
Cameroon
Uganda
Zambia
Mozambique
Senegal
Zimbabwe
Gabon
Botswana
Namibia
South Sudan
Chad
Mauritius
Burkina Faso
Mali
Equatorial Guinea
Madagascar
Congo
Rwanda
Benin
Niger
Guinea
Malawi
Mauritania
Sierra Leone
Eritrea
Swaziland
Togo
Burundi
Lesotho
Liberia
Djibouti
Cape Verde
C. African Republic
Seychelles
Guinea-Bissau
Gambia
Comoros
São Tomé/Prícipe
How many African countries can finance a $1 billion plant?
Source: https://en.wikipedia.org/wiki/list_of_african_countries_by_gdp_(nominal)

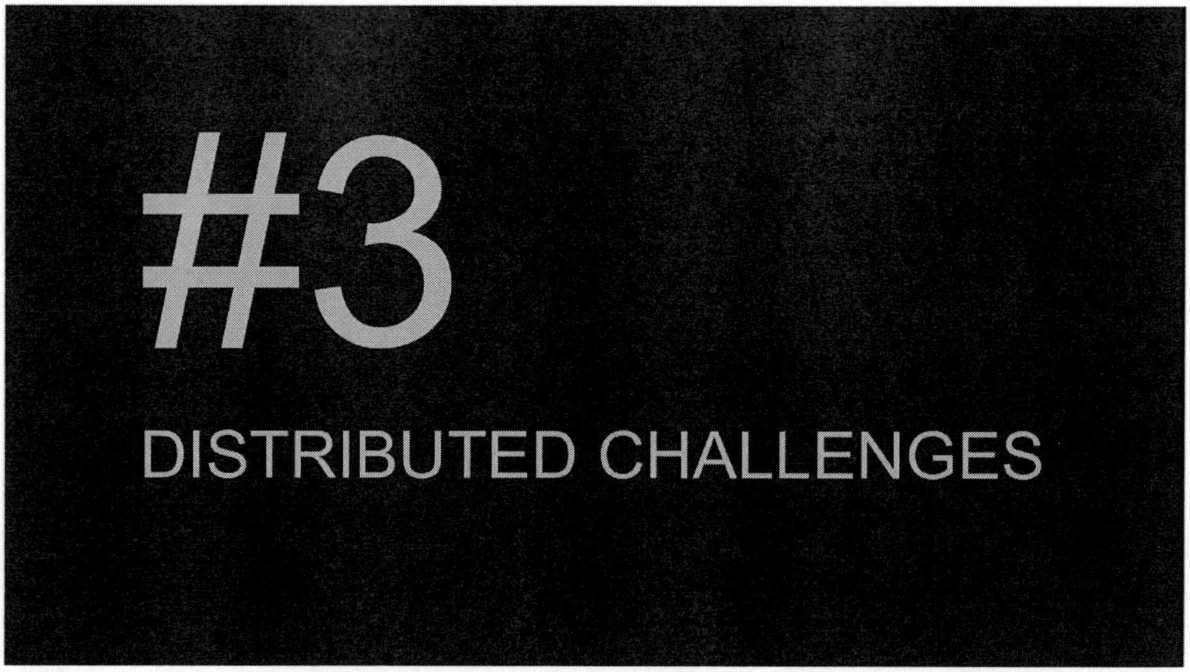

#3
DISTRIBUTED CHALLENGES

Developing World

DISTRIBUTED MARKETS

AFRICA
17% WORLD POPULATION
1% WORLD FERTILIZER USE

In 1999, Uganda farmers bought urea for $600/ton, global market price was $100/ton, why?

- Market size (< 1% global market)
- Transport cost (>$50/ton, 30% total)
- Finance cost ($300k, 1 kton)

Source: World Bank, 2016

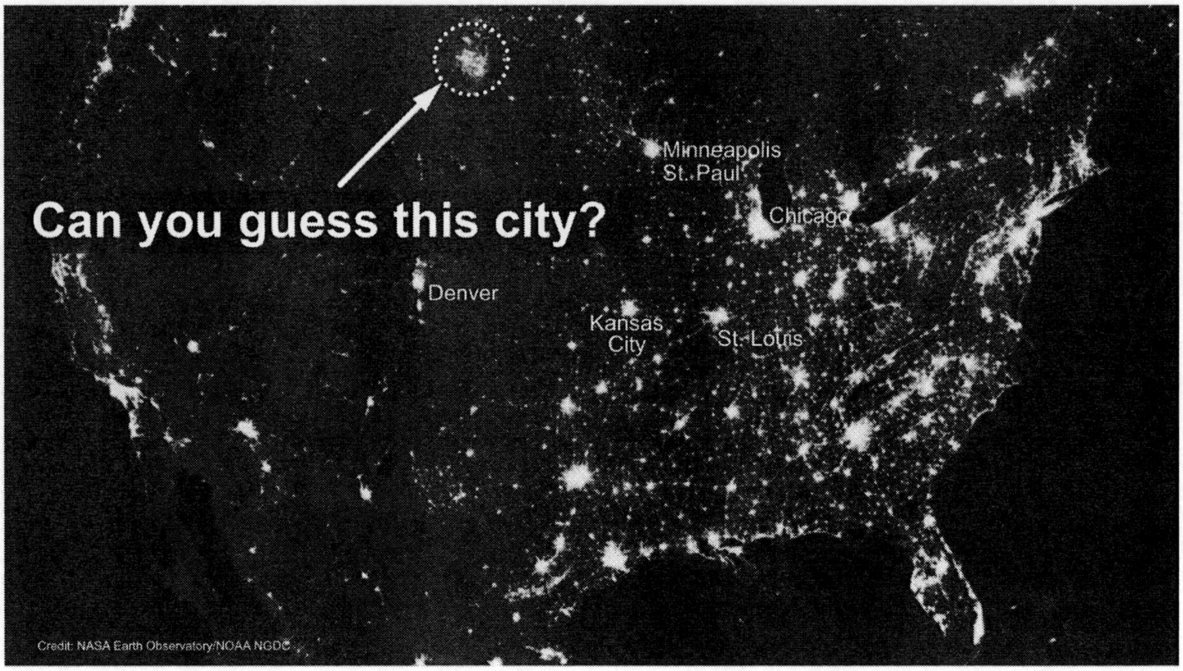

Can you guess this city?
Minneapolis
St. Paul
Chicago
Denver
Kansas City
St. Louis
Credit: NASA Earth Observatory/NOAA NGDC

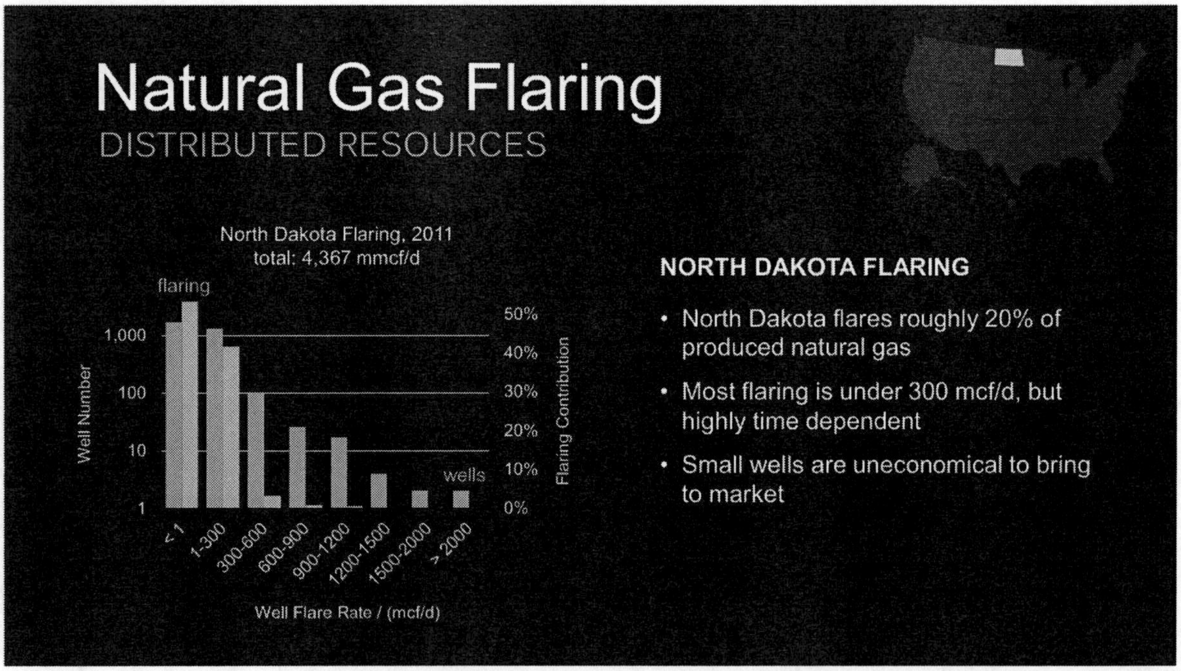

Natural Gas Flaring
DISTRIBUTED RESOURCES
North Dakota Flaring, 2011
total: 4,367 mmcf/d
flaring
wells
Well Number
1,000
100
10
1
50%
40%
30%
20%
10%
0%
Flaring Contribution
< 1
1-300
300-600
600-900
900-1200
1200-1500
1500-2000
> 2000
Well Flare Rate / (mcf/d)
NORTH DAKOTA FLARING
• North Dakota flares roughly 20% of produced natural gas
• Most flaring is under 300 mcf/d, but highly time dependent
• Small wells are uneconomical to bring to market

How do we break the hegemony of scale-up?

Democratizing Technology

HISTORIC EXAMPLES

1450	1908	1950	1973	1977
Gütenburg Press	Ford Model T	ISO Shipping Container	Motorola DynaTAC 8000X	Commodore PET
INFORMATION	TRANSPORTATION	SHIPPING	COMMUNICATION	COMPUTATION

Democratizing Technology

WHAT ARE THE DEFINING FEATURES?

1. Empowers the little guy
2. Never developed by the incumbent
3. Displaces entrenched incumbent
4. Levels the playing field
5. Leverages capital of the many
6. Modular, mass produced, standardized
7. Fast innovation cycles

…and there are many examples

guns	*…displaced*	*swords*
Gutenberg press	*…displaced*	*illuminated manuscripts*
iso-containers	*…displaced*	*bulk shipping*
cellular phones	*…displaced*	*land lines*
personal computers	*…displaced*	*central computing*
automobiles	*…displaced*	*horses*
photovoltaics	*…displaced*	*solar thermal*
electric-arc furnaces	*…displaced*	*blast furnaces*
e-commerce	*…displaced*	*store fronts*
televisions	*…displaced*	*theatres*
record players	*…displaced*	*live musicians*

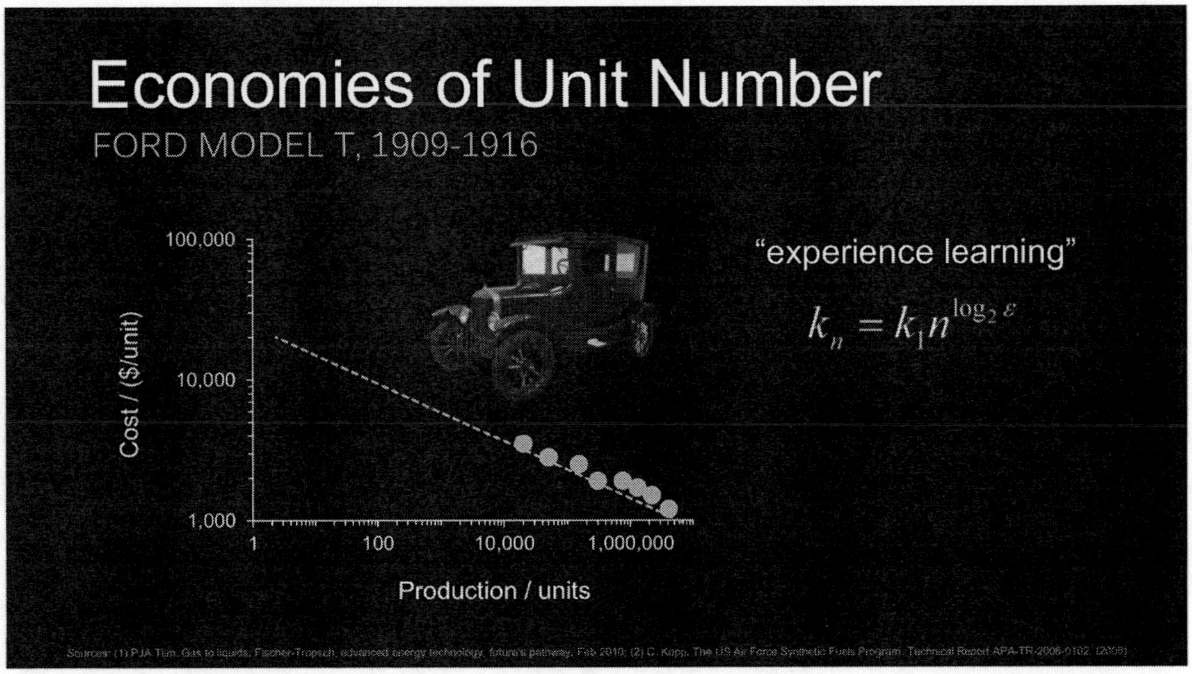
Economies of Unit Number
FORD MODEL T, 1909-1916
100,000
10,000
1,000
Cost / ($/unit)
1
100
10,000
1,000,000
Production / units
"experience learning"
$k_n = k_1 n^{\log_2 \varepsilon}$
Sources: (1) PJA Tijm, Gas to liquids, Fischer-Tropsch, advanced energy technology, future's pathway, Feb 2010; (2) C. Kopp, The US Air Force Synthetic Fuels Program, Technical Report APA-TR-2008-0102 (2008)

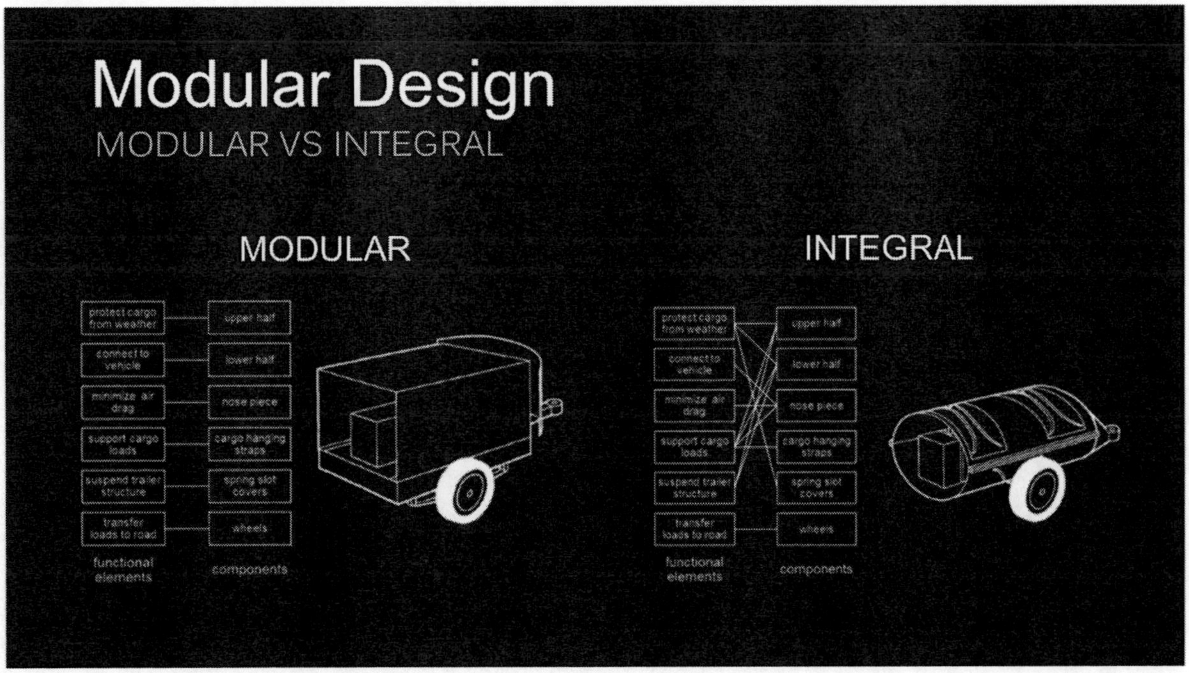
Modular Design
MODULAR VS INTEGRAL
MODULAR
protect cargo from weather
connect to vehicle
minimize air drag
support cargo loads
suspend trailer structure
transfer loads to road
functional elements
upper half
lower half
nose piece
cargo hanging straps
spring slot covers
wheels
components
INTEGRAL
protect cargo from weather
connect to vehicle
minimize air drag
support cargo loads
suspend trailer structure
transfer loads to road
functional elements
upper half
lower half
nose piece
cargo hanging straps
spring slot covers
wheels
components

Modular Design

MODULAR VS INTEGRAL

MODULAR

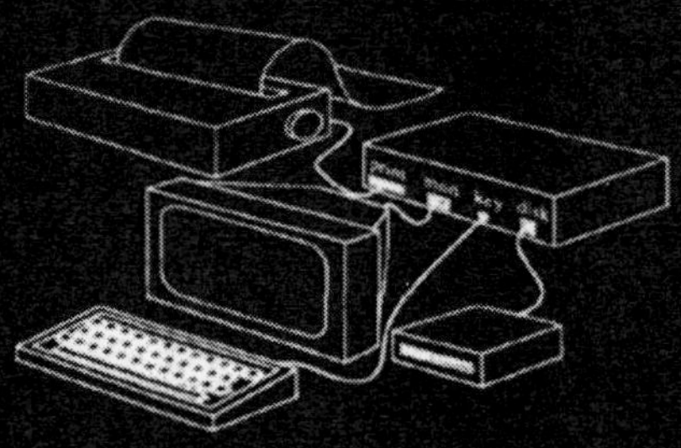

INTEGRAL

Small Modular

OIL REFINERY (WORLD)

- 7 TW petroleum refining
- 700 plants
- $500/kW capex

AUTO ENGINES (U.S.)

- 1.3 TW motive power
- 250 million engines
- $50/kW capex

Small Modular

GAS TURBINES (U.S.)

- 0.2 TW electricity
- 5000 gas turbine generators
- $1000/kW capex

AUTO ENGINES (U.S.)

- 1.3 TW motive power
- 250 million engines
- $50/kW capex

What about the democratization of chemical processing?

not if, but
when, who, how

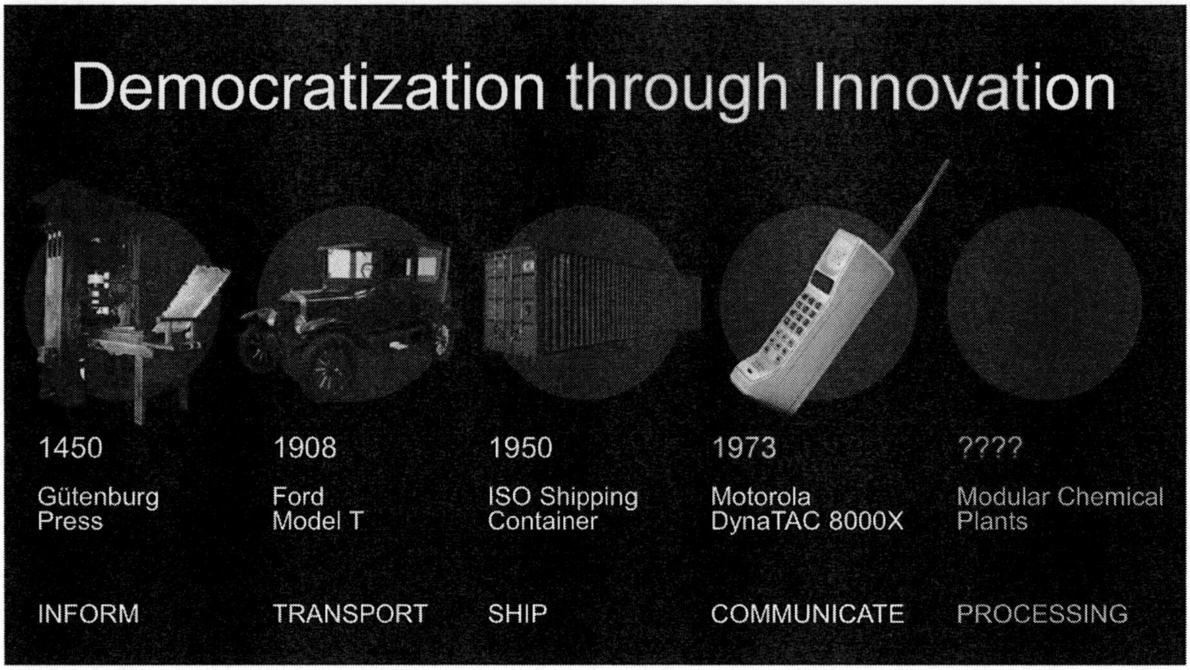
Democratization through Innovation
1450
Gütenburg
Press
INFORM
1908
Ford
Model T
TRANSPORT
1950
ISO Shipping
Container
SHIP
1973
Motorola
DynaTAC 8000X
COMMUNICATE
????
Modular Chemical
Plants
PROCESSING

Isaac Mkalia, 20, checks his mobile phone in Kojiado district, near the
(Photograph: Sven Torfinn/Oxfam)

Thank You

Dane A. Boysen
dane.boysen@gmail.com

505990 The Potential and Challenges of Multifunctional 3D Printing

Abstract:

3D printing has historically been used to produce parts that primarily serve a structural function. However, the potential exists to 3D print blends of materials that serve mechanical, electrical, thermal, optical, magnetic, chemical or other functions. Despite the conceptual appeal of multifunctional 3D printing from an engineering design standpoint, blending polymers, metals, ceramics, and other materials within a part is not trivial. This talk will present methods for 3D printing and curing structures having materials with significantly different thermal processing temperatures. Selected demonstration applications will also be presented.

505811 Applications of Desktop 3D Printing in the Biopharmaceutical Industry

Abstract:

Additive manufacturing (AM), also referred to as 3D printing (3DP), is poised to bring about a revolution in the way drug products are designed, manufactured, and distributed to patients. This technology has gained significant academic acceptance as well as industrial interest and commercial implementation from the ability to create complex geometries along with customizable material properties. Here we demonstrate the use of two common desktop 3D printing approaches, Fused Deposition Modeling (FDM) and Stereolithography (SLA), to manufacture oral and non-oral dosage forms.

Using a re-engineered desktop FDM we succuessfully manufacture thin-walled capsules out of pharmaceutically acceptable materials and co-print with API containing formulations such as liquids, gels, pastes and powders. The main advantage to this approach is our ability to design and create custom API release profiles to test in a clinical setting. To date we have developed three distinct designs that provide unique and informative dissolution profiles: 1) osmotic-like zero-order releases using nondissolvable capsules with dissolvable capsule features, 2) controlled delay releases using variable thickness, and 3) pulsatile releases using pill-in-a-pill designs. Additionally, the agility in this approach potentially introduces a rapid on-site manufacturing for personalized dosing.

For non-oral applications, we demonstrate a simple, low cost, and commercially available method for fabricating microneedle arrays using a desktop SLA printer. The default settings of the printer introduce defects into the fabricated microneedles, but through proper optimization, using a combination of reducing layer height, employing a high quality antialiasing algorithm, and rescaling the input images enables the production of high fidelity microneedles. These microneedles demonstrate sharp tip radii with fabrication times less than one hour. Further, we demonstrate that the opimal microneedle design is accomplished through adjustments in the height, width, and spacing of these arrays. Microneedle masters produced using this technique is combined with standard micromolding approaches to fabricate microneedles from desired materials, such as water soluble or biodegradable polymers. We anticipate that this approach will lower the barrier to entry into the microneedle field for researchers with little background in microfabrication and provide an easy way to adjust key microneedle parameters, such as size, aspect ratio, and spacing tuned for the active ingredient and drug delivery rate.

505750 Three-Dimensional Printing via "Capillary Engineering" of Multiphasic Elastomer Inks

Abstract:

The talk will present a new method for 3D printing with soft silicone-based material by using inks made from multiphasic liquid/liquid/solid systems. These inks consist of water and two components, crosslinked polydimethylsiloxane (PDMS) microbeads and a liquid non-crosslinked PDMS phase. Owing to the capillary binding of the microbeads, their suspensions containing certain fractions of liquid PDMS precursor behave like thixotropic pastes, which are flowable at high shear stress, yet they possess high storage moduli and yield stresses needed for direct ink writing. Their rheology can be controlled by the volume fraction of the beads and liquid binder. These PDMS gels can be directly extruded and shaped on a 3D printer. The liquid bridges are thermally crosslinked after printing, resulting in remarkably elastic and flexible PDMS structures, due to the chemical similarity of the solid beads and the liquid phase that has been bridging them. Their porosity and mechanical properties, such as tensile modulus, could be controlled by the fraction of liquid precursor in the original multiphasic dispersion. The high softness, elasticity, and resilience of these 3D printed PDMS materials open new opportunities in the making of highly flexible 2D meshes and 3D macrostructures (patent appl. 2016, *Adv. Mater.* 2017). As this ink is made of porous biocompatible silicone and can be 3D printed under water, it may also find applications such as direct printing of bio-scaffolds on live tissue.

506189 Advancing Diagnostics and Therapies by Enabling Scientists with 3D-Printing Technologies

Abstract:

The Spence group, housed in the new Institute for Quantitative Health Science and Engineering at Michigan State University, currently has 3 core projects under investigation. Broadly defined, these projects are in the fields of (1) diabetes, (2) multiple sclerosis (MS), and (3) blood banking. While a number of quantitative measurement schemes (classic scintillation counting, microscopy, mass spectrometry) and methods (ELISA, cell culture, separation science) are used to investigate these problems, it has been our ability to fabricate 3D-printed devices that has advanced our understanding of the molecular level events important to each project. Various enabling technologies from the 3D-printer will be shown throughout the talk, with an emphasis on 3D-printed, *in vitro* devices that either enable cell to cell communication or, as in the case with our blood banking efforts, provide scaled-down versions of protocols for studies in our academic laboratories. Importantly, many of these prototyping efforts were not possible with other â €œacademicâ€ prototyping methods based on soft-lithographic procedures, a mainstay in our group for nearly a decade.

507189 Perfusion Directed 3D Bone Mineralization Using Custom-Modified 3D Printers

Abstract:

An inherent challenge in conventional tissue engineering strategies is the ability to efficiently deliver nutrients throughout the thickness of a complex, physiologically relevant biomimetic construct. In this work, we will present the creation of structurally supported, perfusable hydrogels capable of growing bone in user defined directions. Briefly, bone-like human osteosarcoma cells were encapsulated inside UV cross-linkable gelatin methacrylate (GelMA) hydrogels, and this cell-hydrogel mixture was casted onto a 3D printed poly(vinyl alcohol) (PVA) structure. PVA serves as a sacrificial material and was dissolved away to obtain hollow channels to facilitate the perfusion of media using a custom-made acrylonitrile butadiene styrene (ABS) bioreactor. Osteogenic media was perfused through the channels, and the radial zones of bone mineralization surrounding the channels were quantified. This study demonstrates that user-defined 3D printed channels can be used to spatially control bone mineralization.

502794 Vapor Recompression Mechanism Introducing in Internal Heat-Integrated Distillation Column: Impact of Internal Energy Driven Intermediate and Bottom Reboiler

Abstract:

A novel combination of internally heat-integrated distillation column (HIDiC) and vapor recompression column (VRC) with intermediate reboiler (IR) is proposed. Supplying heat at the highest temperature point (i.e., column bottom) of the VRC scheme is not thermodynamically favorable and, therefore, we aim to install the IR for better distribution of heat along the column length, thereby reducing the compressor work. Introducing IR in the combined HIDiC-VRC system formulates an open-loop variable manipulation policy to evaluate the comparative impact of internal and external heat sources on bottom liquid reboiling. With internal energy driven bottom reboiler, we further investigate the hybrid HIDiCVRCIR column with proposing the two modes of compressor arrangement, namely parallel and series. Finally, a multicomponent distillation system is exampled to show the promising potential of the proposed HIDiC-VRCIR configurations in improving the energetic and economic performance over the HIDiC-alone and HIDiC-VRC schemes with reference to a conventional standalone column.

496355 Experimental Studies Controlling Trace Components in a Dividing Wall Distillation Column

Abstract:

Experimental Studies Controlling Trace Components in a Dividing Wall Distillation Column

Melissa Donahue, Michael Baldea, R. Bruce Eldridge

McKetta Department of Chemical Engineering and

James R. Fair Process Science and Technology Center

The University of Texas at Austin

An example of one of the prominent technologies in the field of process intensification is dividing wall distillation columns or DWCs. As part of an ongoing experimental study of dividing wall columns at The University of Texas at Austin[1], the design and selection of control strategies for a pilot plant DWC will be presented. The pilot plant column is six inches in diameter and approximately 30 feet tall. The column has 24 theoretical stages of structured packing and a welded wall in the horizontal and vertical middle of the column[1]. To investigate the operational flexibility of this fixed design DWC, a fourth trace component is added to the feed. A sensitivity analysis studying the minimum energy usage and optimal distribution of impurities is conducted using a dynamic process simulator. These operating targets and conventional distillation temperature control techniques are used to design control configurations that are then implemented on the pilot scale dividing wall column. The control configurations are tested against changes in set point and feed flow and composition disturbances.

(1) Blevins, T.; Downs, J. J.; Donahue, M. M.; Roach, B. J. Use Model Predictive Control to Achieve Real-Time Management of a DWC. *Hydrocarbon Processing* **2015**.

CASCADE REACTION IN A SPINNING MESH DISC REACTOR USING A NOVEL METAL- ENZYME CATALYST

PARIMALA SHIVAPRASAD
Department of Chemical Engineering
ps784@bath.ac.uk

Emma Emanuelsson*, Darrell Patterson*,Matthew Jones**
* Department of Chemical Engineering, University of Bath
** Department of Chemistry, University of Bath

Bath - A University in a UNESCO World Heritage City

Outline

UNIVERSITY OF BATH

Background

The Idea

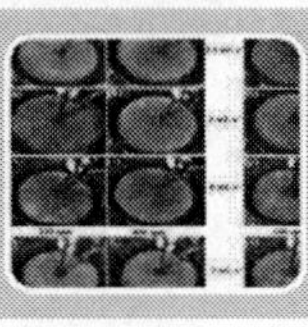

Results

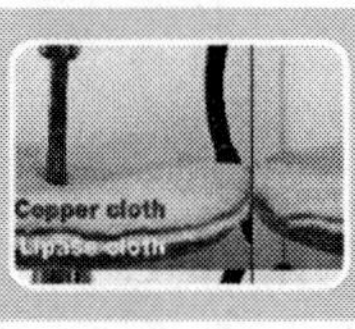

The Path Forward

Problem

UNIVERSITY OF BATH

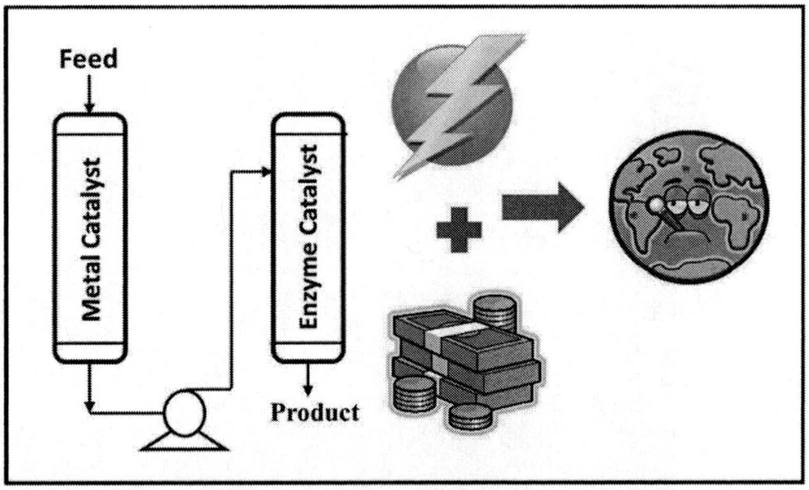

Multi-step processes lead to higher recovery and energy costs.

Solution

UNIVERSITY OF BATH

Imagine a reactor that could do a cascade of reactions without the need for product purification in between!

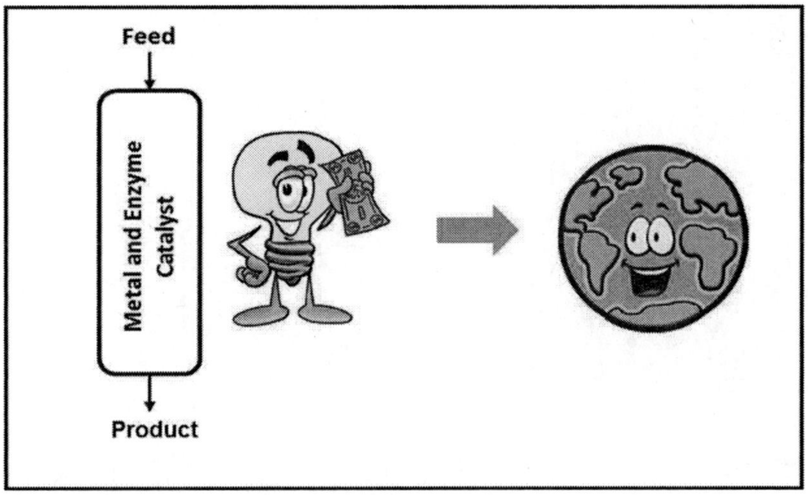

Catalyst immobilisation on wool

UNIVERSITY OF BATH

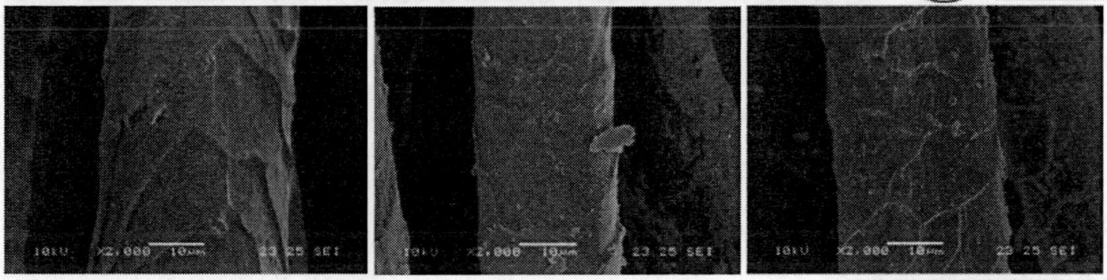

Shivaprasad, P., *et al.(2017) Chemical Engineering and Processing: Process Intensification.*

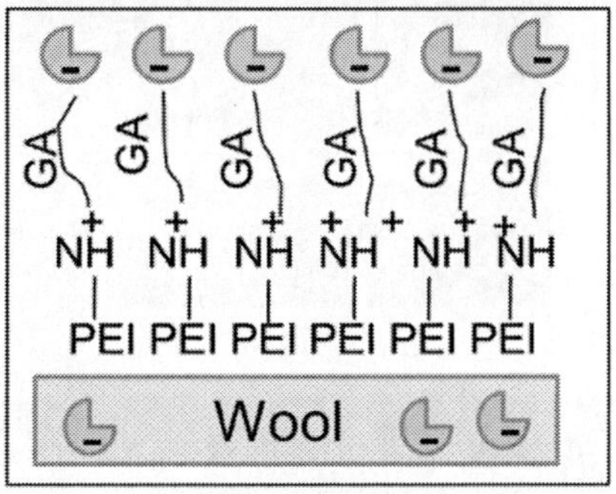

Feng, X., *et al.(2012) Colloids and Surfaces B: Biointerfaces*, 102, pp. 526-533

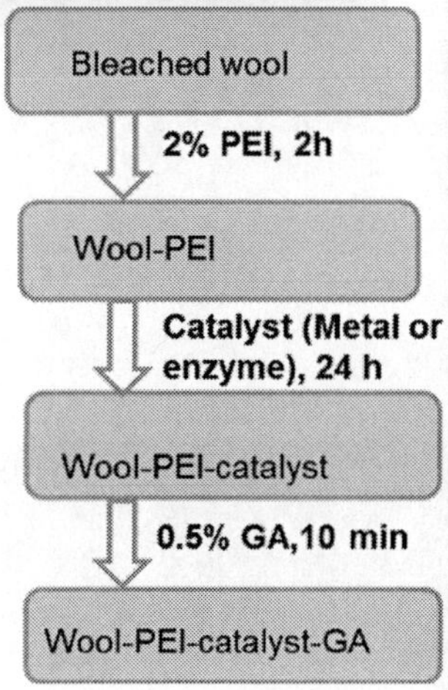

The Idea: Process Intensification

UNIVERSITY OF BATH

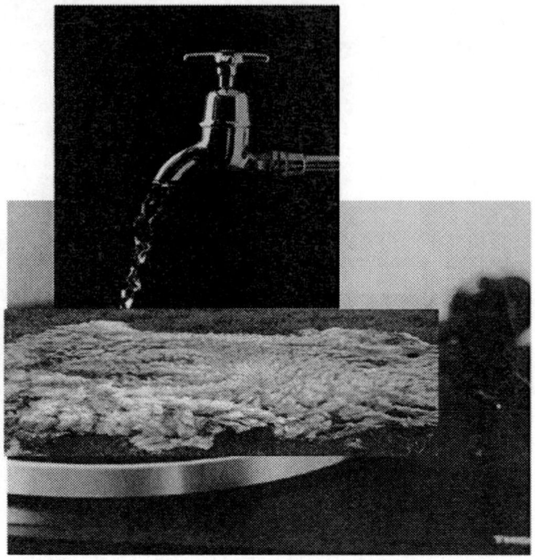

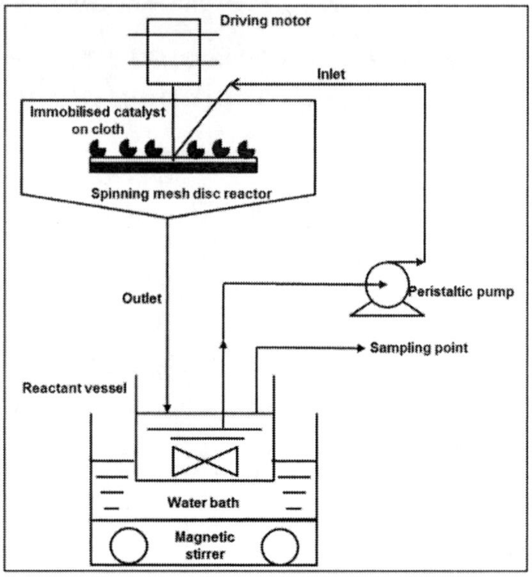

Spinning Mesh Disc Reactor (SMDR)

The SMDR: Characteristics

UNIVERSITY OF BATH

Reactor characterisation for enzymatic hydrolysis of tributyrin

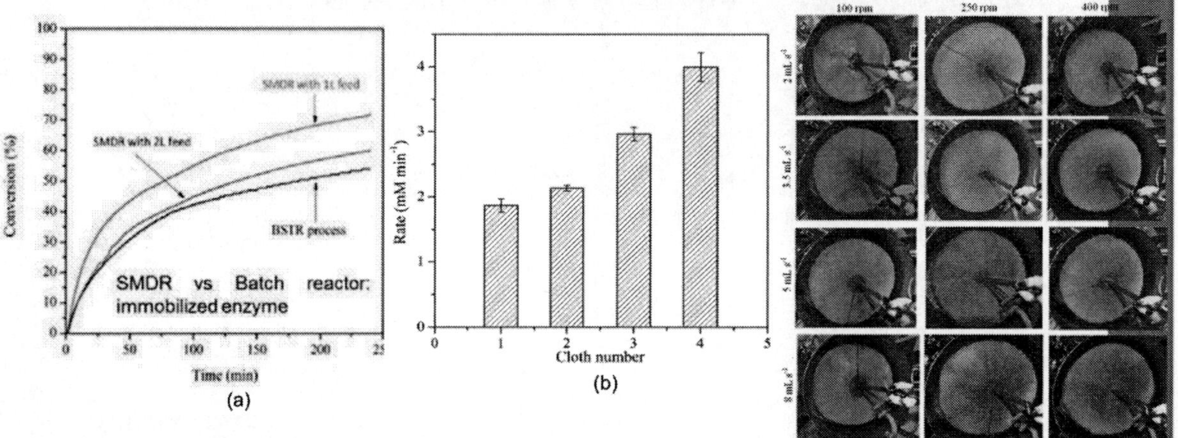

Spinning Mesh Disc Reactor

- ✓ Process Intensification
- ✓ High Reusability of enzyme
- ✓ Scale-up

Feng *et al.* (2013) *Chemical Engineering Journal*, 221, 407-417

Towards a one pot cascade reaction

UNIVERSITY OF BATH

Dynamic Kinetic Resolution (DKR) of Henry Reaction

Perfecting individual reactions in the SMDR

STEP 1: Copper catalysed Henry reaction

STEP 2: Enzyme catalysed kinetic resolution of racemic alcohol

One-pot cascade reaction

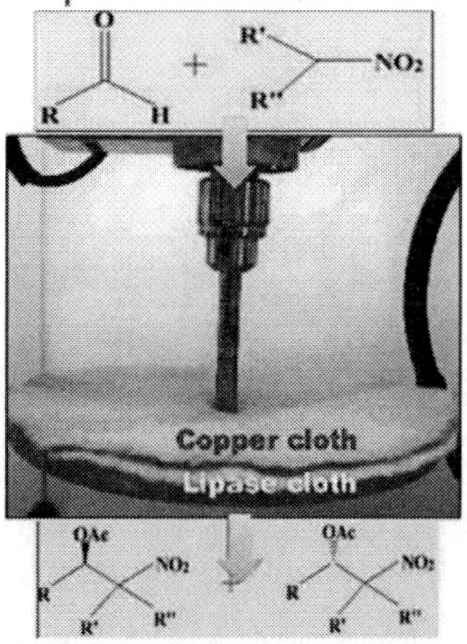

Copper-lipase catalyst in tandem

The Henry reaction

UNIVERSITY OF BATH

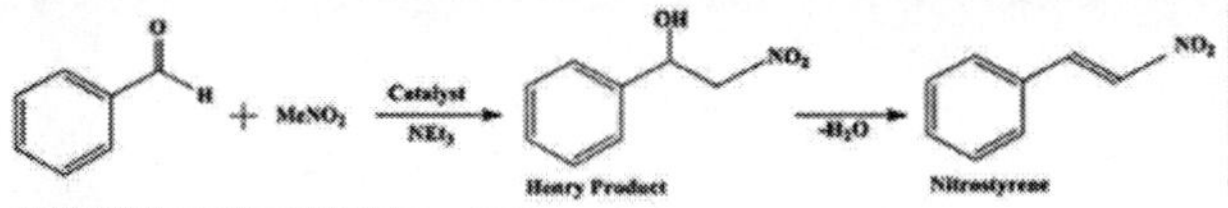

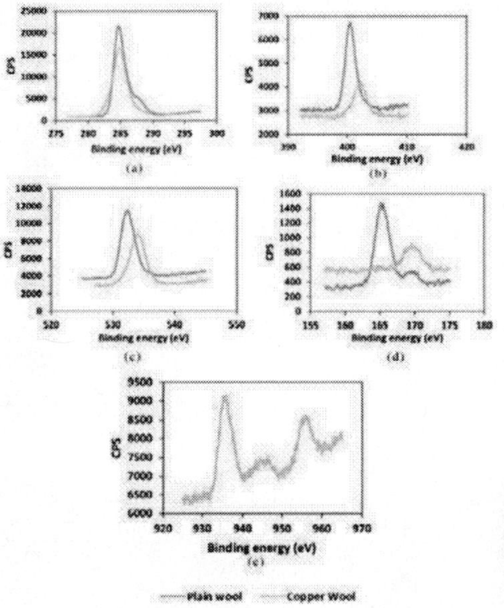

Process intensification of catalysed henry reaction using copper-wool catalyst in a spinning mesh disc reactor

Parimala Shivaprasad, Matthew David Jones, Darrell Alec Patterson, Emma Anna Carolina Emanuelsson

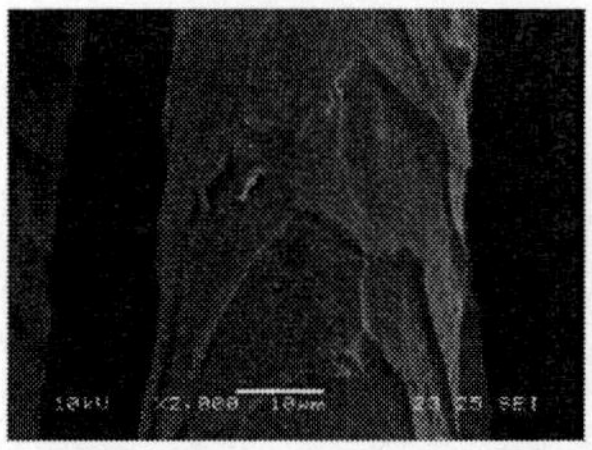

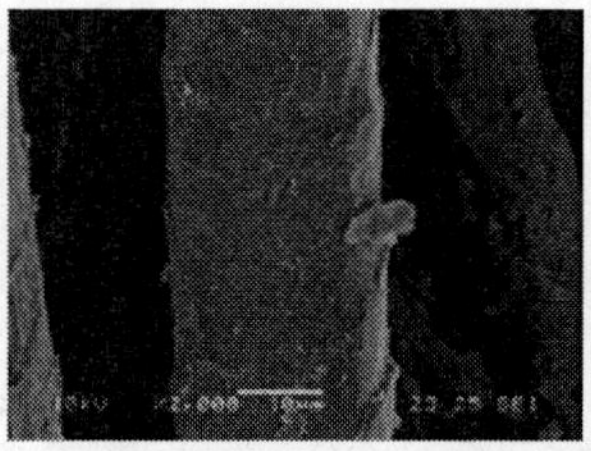

Copper triflate was successfully immobilised on wool

The Henry reaction

UNIVERSITY OF BATH

Henry reaction with free and immobilised catalyst in batch

Catalyst	Time (h)	% Conversion
Free copper triflate	4	38
(homogeneous batch)	6	42
	24	57
Copper cloth	4	41
(heterogeneous batch)	6	63
	24	85

Henry reaction using the copper cloth in the SMDR

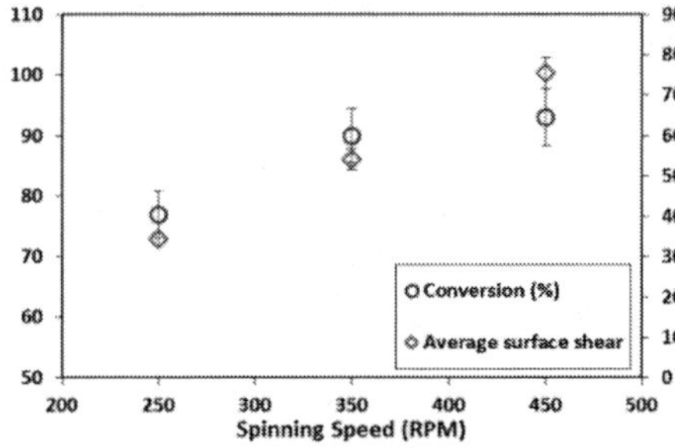

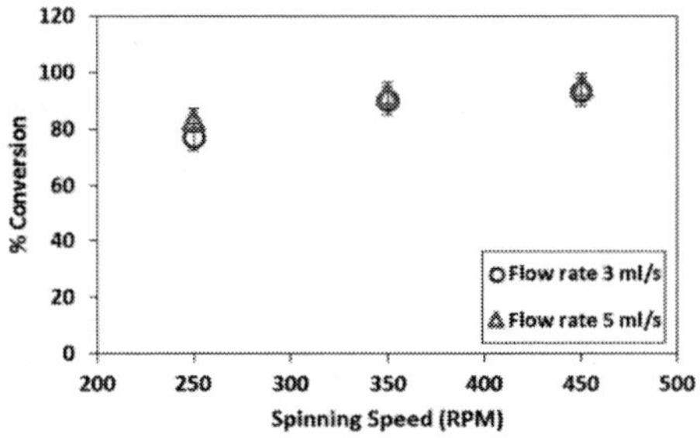

- ✓ Copper immobilised on wool showed a higher conversion in batch.
- ✓ Higher reaction rate was observed in the SMDR using the copper cloth
- ✓ The SMDR was successfully applied for a solvent based reaction.

Shivaprasad, P., *et al.*(2017) *Chemical Engineering and Processing: Process Intensification.*

Wool as a catalyst

UNIVERSITY OF BATH

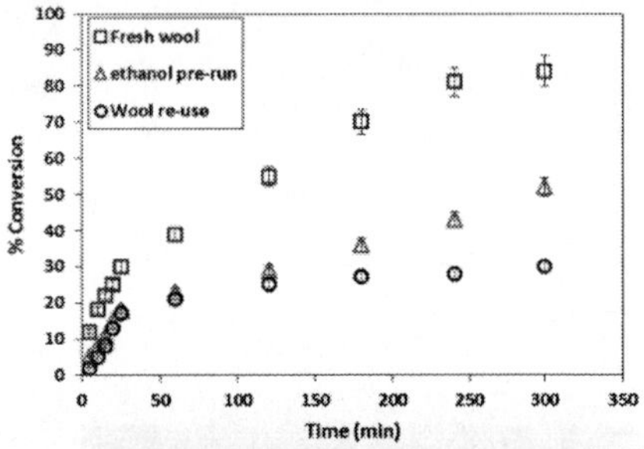

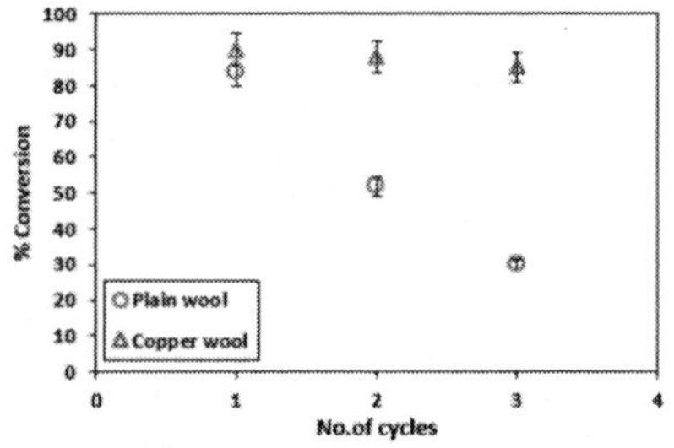

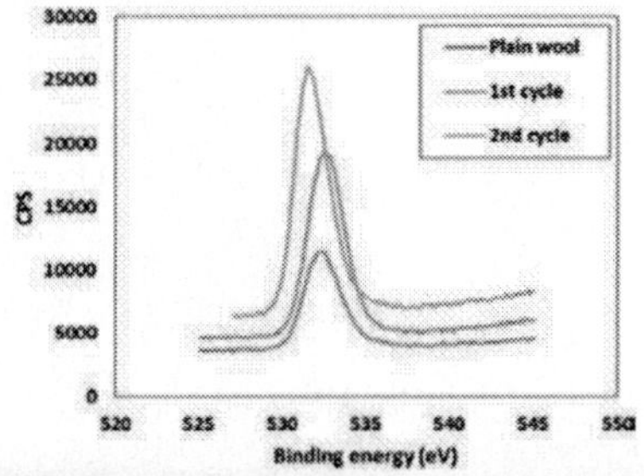

- ✓ Wool was found to catalyse the Henry reaction but was not re-usable for multiple cycles.
- ✓ Copper triflate immobilised on wool was more robust than plain wool.

Shivaprasad, P., *et al.*(2017) *Chemical Engineering and Processing: Process Intensification.*

Enzymatic kinetic resolution in the SMDR

UNIVERSITY OF BATH

Kinetic resolution of 1-phenyl ethanol catalysed by lipase

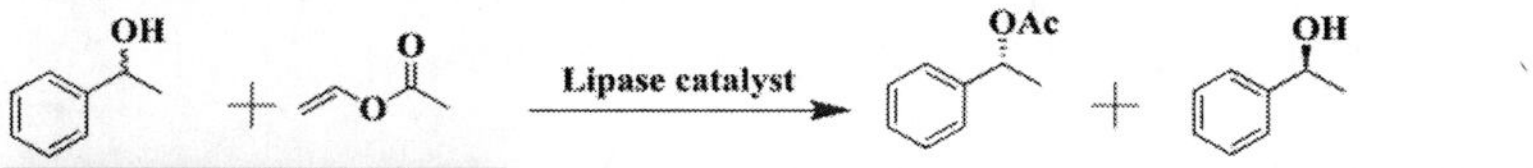

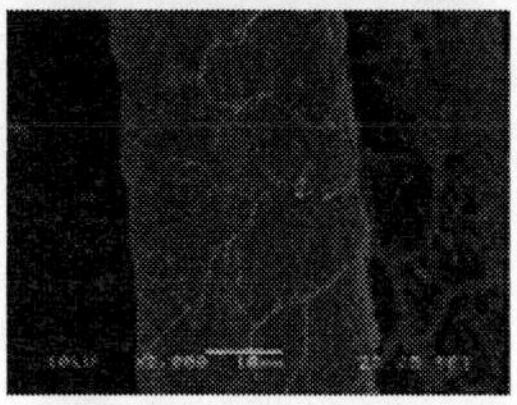

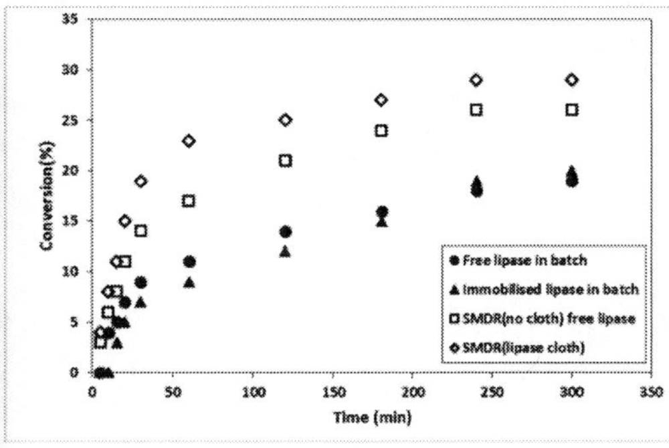

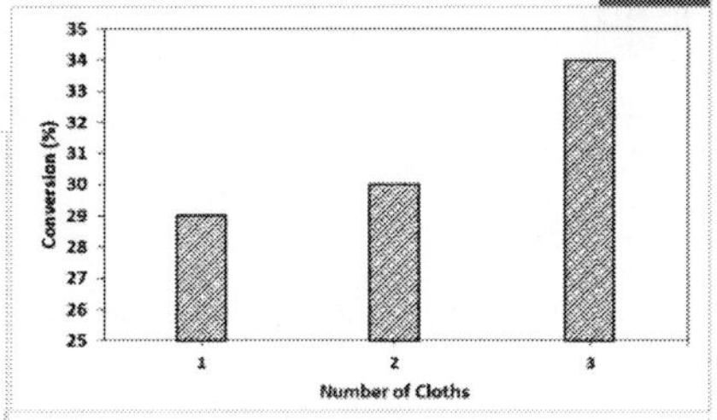

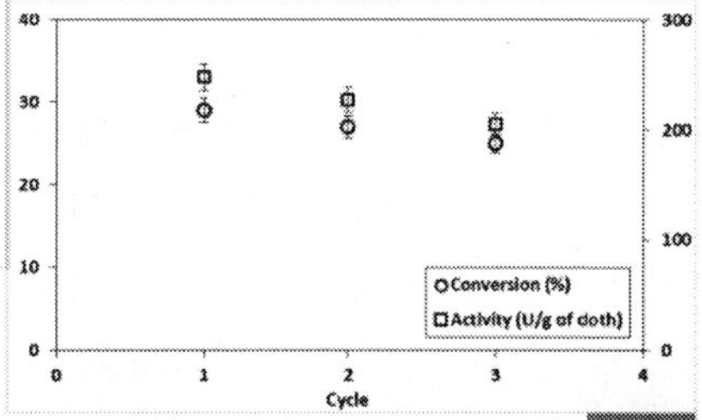

- ✓ Lipase immobilise on wool was used for the KR reaction in both batch and SMDR
- ✓ The reaction rate in the SMDR was higher for both free and immobilised lipase compared to the batch reaction.

TOWARDS CASCADE REACTIONS

UNIVERSITY OF BATH

Can copper and lipase immobilised on wool catalyse the reaction in tandem?

Challenges

- ✓ Solvent compatibility
- ✓ Robustness of the metal and the enzyme catalyst in one-pot
- ✓ Reaction kinetics

SUMMARY

UNIVERSITY OF BATH

Process intensification compared to more conventional reactors under equivalent operating conditions:

- enhanced reaction rate
- enhanced enzyme stability

The SMDR was used to demonstrate traditional organic synthesis reactions:

- The Henry reaction
- Enzyme catalysed kinetic resolution

Future work

UNIVERSITY OF BATH

One-Pot Cascade reaction

- Different enzyme and catalyst

Spinning Mesh Disc Reactor – further characterisation

- Increasing cloth size vs stacking cloths?
- A prototype for 'Perfect Reactor' – The future of process intensification

Acknowledgement

UNIVERSITY OF BATH

- Emanuelsson and Patterson Research Group
- Dr. Xudong Feng and Kemi Lawal
- The Santander mobility fund
- University of Bath and European Union's Seventh Framework Programme for research funding.

QUESTIONS?

484756 Scale and Effects of Catalyst Deactivation in Enzymatic Catalyzed Reactive Distillation

Abstract:

Increasing global competition and environmental legislation require the chemical sector to produce more cost efficient and reduce emissions. Process integration reduces the energy consumption of a process as well as its investment costs. An example for an integrated process is the combination of reaction and separation in form of the reactive distillation (RD) or reactive dividing wall column (RDWC). However, many of the industrially employed catalysts also enhance the production of unwanted side products that require additional, expensive purification steps. The employment of enzymatic catalysts is a promising approach to greatly increase the selectivity and reduce the required temperatures for chemical reactions. As enzymes are often product inhibited, the in situ product removal of a reactive distillation allows fully taking advantage of the enzymeâ€™s potential. Studies on laboratory and pilot plant scale columns have already shown the feasibility of this approach[1,2]. For the next step, the industrial application in a continuous reactive distillation, the stability of the enzymes is a fundamental condition, but yet untested under operating conditions. Therefore, knowledge about the scale of deactivation, the influencing factors and how to operate a column to minimize deactivation and maximize the production time before a standstill is required, are key questions still to be answered.

Our research aims to systematically investigate the behavior of eRDWCs. The first step has been to show the general feasibility of a successful operation with a continuous eRDWC 12 m high pilot plant.[1] Focus of this work was to develop a start-up strategy and to show that the column can be operated successfully for different operating conditions. Using the acquired experimental data a comprehensive simulation model was validated. The next step is a comprehensive approach to answer the questions about enzyme stability that hinder the industrial application. To analyze the scale and driving forces of enzyme deactivation two different experimental set-ups have been built. First, the eRDWC pilot plant and additionally a fixed-bed reactor have been employed in long term studies to test enzyme deactivation. After examining the experimental results on the quantity and driving forces of deactivation the second part of the project uses the developed mathematical model to analyze the influence on deactivation. The combined results from experimental and simulation studies allow a much deeper understanding about how to minimize enzyme deactivation in reactive distillation and additionally how to achieve the required conversion and purities for reduced enzyme activities.

This presentation will give a comprehensive overview about the scale and underlying causes of enzyme deactivation. A short introduction will explain how an eRDWC works and when to employ it. The simulation model and pilot plant will be presented. Selected experimental results about enzyme activity from the eRDWC and fixed-bed reactor experiments will be shown. From this data conclusions will be drawn about scale and influencing factors. Finally, it will be explained how to limit the enzyme deactivation by employing a smart process control and how to maintain the product specifications if deactivation occurs.

References

1.Egger T, Fieg G. Enzymatic catalyzed reactive dividing wall column: Experiments and model validation. AIChE J. 2016. doi:10.1002/aic.15598 .

2.Heils R, Jensen J, Wichert S, et al. Enzymatic Reactive Distillation: Kinetic Resolution of rac -2-Pentanol with Biocatalytic Coatings on Structured Packings. Ind. Eng. Chem. Res. 2015;54(38):9458-9467. doi:10.1021/acs.iecr.5b02802 .

498215 Use of Computational Fluid Dynamics (CFD) Simulation and Image Analysis Tool for Modelling Light in a Microalgal Photobioreactor

Abstract:

Light, one of the most important parameter for the growth of microalgal cells, is converted into chemical energy by the process of oxygenic photosynthesis. Microalga is a valuable feedstock for the production of industrially important metabolites. The energy for algal growth is obtained from the visible light within the photosynthetically active radiation (PAR) ranging from 400 to 700 nm wavelength. The high density of microalgal cultures is usually attained by cultivation in the closed photo-bioreactors (PBR) as compared to open systems due to better process control, mixing, light distribution etc. However, there is a need for optimizing the cultivation processes that should support both autotrophic and mixotrophic cultivation. Light in particular is known to directly influence the algal biomass production and product yield (Kumar, Sirasale, & Das, 2013). The conventional algal PBRs suffers from self-shading effect which is caused when microalga attains high cell density, thereby reducing the light penetration inside PBR and leading to development of dark zones. This is mainly because gradient of incident light intensity varies along the radius of PBR. Thus, local light intensity distribution at any given point inside the PBR is dependent on the path length, light penetration and absorption coefficient that becomes much acute especially during scaling-up of the process. Existing PBR designs do not have the desired efficiencies, and lead to high material and energy costs. Moreover, predicting the hydrodynamics (fluid flow pattern) of a PBR in laboratory experiments is difficult and cumbersome. Therefore, novel PBR designs need to be developed and benchmarked against existing designs.
Light utilization in the microalgal PBRâ€™s is usually measured by incident light intensity on the surface of the PBR or the irradiating surface area per unit culture volume. Average light intensity received by the microalgal cells is used for measuring the light utilization or photosynthetic efficiency of photosynthetic microorganisms. Efficient mixing allows cells to continuously pass through various zones inside a PBR such as complete dark, photo-limiting (cell growth increases with the light intensity), saturating and photo-inhibition (increase in light intensity leads to growth inhibition) and gives them flashing light effect. Despite several simulation studies on PBRs, a comprehensive analysis and evaluation of light penetration in PBRs has been limited.
The present study discusses use of CFD simulation for designing a novel airlift PBR with improved hydrodynamics of fluid flow. The novel PBR will be used for the cultivation of microalgae. The designed reactor performance in terms of light distribution will be compared with the conventional bubble column PBR. Since the application of multiphase CFD is complicated due to difficulty in describing the variety of interactions in these systems. The hydrodynamics of the reactor will be simulated using CFD and later will be used for microalgal cultivation. The present work describes the use of image analysis technique for the measurement and tracking of cell growth at different growth phases of a live bio-culture in the indigenously designed novel PBR. The light penetrability of incident light intensity was assessed in a Chlorella sp. culture throughout a batch experiment. The light penetration profile could be generated following modified Beer-Lambertâ€™s law by spectrophotometric method. Digital images of the top view of the PBRâ€™s were taken and processed using image processing tool in the MATLAB software. This was used to estimate the light intensity distribution in the externally radiating column PBR across the radial path length. Light gradient pattern inside the algal suspension was assessed by combining fluid dynamics simulations of three-dimensional turbulent single-phase fluid flow with statistical particle tracking and signal analysis. (Perner-Nochta & Posten, 2007). The experimental results and CFD simulations exhibited a good fit. Also, the light penetration was mapped with the image analysis tool for a better understanding of light availability and detection of dark zones inside a working PBR.

494356 IGCC Process Intensification with the Reforming Technologies for Enhanced Power Generation with CCS Technology

Abstract:

IGCC (Integrated gasification and combined cycle) is a pre-combustion technique which represents higher thermal efficiency with large scale implementation of CO_2 capture. On the other hand, NGCC (Natural gas combined cycle) power plants offers higher process performance compared to IGCC process with CCS technologies but the fluctuating and comparatively higher costs of natural gas limits its extensive implementation over coal based power generation systems. Therefore, process intensification of coal based gasification processes with the natural gas reforming technology has been proposed in this study to enhance the overall process performance while limiting the natural gas consumption. In this study, two IGCC based process models have been developed and evaluated in terms of both the process performance and economics with CO_2 capture. Case 1 is taken as the conventional IGCC process, whereas, case 2 presents an idea of integrating methane reforming process with an IGCC technology. Coal is used as primary fuel in both the cases for the coal gasification process, however, small amount of natural gas is used as secondary fuel only in the case 2 design. The high enthalpy steam generated during coal slurry gasification process in case 2 is used to assist the methane reforming to generate additional H_2. The integration of IGCC with methane reforming process not only supplies the heat required for the endothermic reforming process but also increases the heating value of the resulting syngas. Moreover, the H_2O generated in the gasification process in case 2 can be consequently utilized in the natural gas reforming process without utilizing any extra steam. This concept also provides an opportunity for the process economics since shared water gas shift reactors and CO_2 capture units will suffice the process needs. The net electricity generation capacity and efficiency for case 1 and case 2 is calculated as (375.08 MW_e, 472.92 MW_e) and (35.93%, 40.70%), respectively. While comparing the results, it has been seen that case 2 design offers nearly 4.77% higher efficiency compared case 1 design with CO_2 capture. The process economics analysis showed that the case 2 design requires a higher CAPEX and OPEX throughout the project life compared to the case 1 design. However, due to the higher power generation capacity of case 2 design, it has a potential to reduce the LCOE (levelized cost of electricity) by 5.81% compared to the case 1 design. Moreover, case 2 design not only requires the less CO_2 capture and avoidance costs compared to the case 1 but also exhibits higher rate of return on the investment with the less payback time. With the higher efficiency and least case 2 has been considered as the most feasible option for electricity production at an economical price compared to the case 1 design.

500239 Vegetable Oil Process Intensification with Sterically Hindered Alcohols to Biofuels and Biochemicals

Abstract:

Vegetable oils, refined to different degrees, are the raw material to manufacture biodiesel. Commercial batch times can exceed 8 h due to poor mass transfer and miscibility between oil and methanol. Moreover, free fatty acids methyl esters (FAME) have a cloud point (CP) and pour point (PP) ranging from −3 °C to 12 °C and −15 °C to 10 °C, respectively [1], i.e. poor cold flow properties unsuitable for most climates unless blended with petro-diesel. Substituting methanol with branched alcohols to transesterify or esterify vegetable oil matrices improves the cold flow properties of the fatty alkyl esters. For example, the CP and PP of isopropyl esters range from −14 °C to 0 °C and −20 °C to −5 °C, respectively [1]. We propose a fluidized catalytic bed and ultrasound-based processes to either transesterify vegetable oils with alcohols other than methanol or esterify free fatty acids with a trimethylolpropane to produce biolubricants.

A fluidized catalytic bed concurrently transesterified and cracked canola oil with iso-propanol (iPrOH) to fatty acid isopropyl esters in the vapour phase. We alternated the injection of vegetable oil + iPrOH (reaction stage) and air (regeneration stage) to remove the coke from the CaO/Al_2O_3 catalyst. The reaction stages and regeneration stages lasted 2 min. We varied the iPrOH:oil ratio from 35 to 51 (mol:mol) and the air flow for during the regeneration step. The coke yield ranged from 3.5% to 34% and the isopropyl esters yield from 2% to 31%. The highest hydrocarbons yield was 42% (GC analysis). Fluidized beds are excellent candidates for biodiesel process intensification because as little as 10 t of catalyst may produce 80 kt/y of biodiesel, which corresponds to a reactor size at least 10X lower than the current commercial processes [2-4].

We propose a new ultrasonic-assisted mixing device to transesterify vegetable and waste cooking oil with ethanol and iso-propanol in a continuous process. We adapted a 13 mm diameter tip ultrasound probe to a media mill, where a rotating basket circulated the oil: alcohol mixture at 1200 rpm in the vicinity of the tip [5]. The alcohol:oil ratio was stoichiometric and we adopted KOH as basic catalyst. Ultrasound operated at 20 KHz and 500 W nominal power and with pulses 1 s on and 1 s off. After 1 min reaction, 40 %, 55 % and 90 % oil converted to isopropyl, ethyl and methyl esters, respectively. In absence of ultrasound, the same conversions were reached after 2 hours. Ultrasound creates a very fine oil: alcohol emulsion that increases the contact among the reagents and the catalyst.

Vegetable oils as biolubricants suffer from poor oxidative stability because of the hydrogen in b position to the ester groups, which have the tendency to give b-elimination [6]. Trimethylolpropane (TMP) esterify free fatty acids to biodegradable and stable tri-esters whose application is as biolubricants. We apply ultrasound to esterify oleic acid with TMP with heterogeneous catalysts including Al_2O_3-supported $H_3PW_{12}O_{40}$ and sulphated ZrO_2 systems. We vary the power amplitude (20 to 60 % of 500 W) and adopt different ultrasound pulses lengths. Sulphated zirconia systems convert over 80 % of substrate in 2 hours, while Al_2O_3-supported $H_3PW_{12}O_{40}$ is mostly inactive.

References

[1] Joshi R.M., Pegg M.J., Flow properties of biodiesel fuel blends at low temperatures, *Fuel*, 2007, 86:143–151

[2] Boffito D.C., Galli F., Pirola C., Patience G.S., CaO and isopropanol transesterify and crack triglycerides to isopropyl esters and green diesel, *Energ. Conv. Mangement.*, 2017, 139, 71-78

[3] Boffito D.C., Neagoe C., Edake M., Pastor-Ramirez B., Patience G.S., Biofuel synthesis in a capillary fluidized bed, *Cat. Today*, 2014, 237, 13–17

[4] Boffito D.C., Blanco-Manrique G., Patience G.S., One step cracking/transesterification of vegetable oil: Reaction–regeneration cycles in a capillary fluidized bed, *Energ. Conv. Manage.*, 2015, 103, 958–96

[5] Boffito D.C., Galli F., Martinez P.R., Pirola C., Bianchi C.L., Patience G.S., Transesterification of triglycerides in a new ultrasonic-assisted mixing device, *Chem. Eng. Trans.*, 2015, 43, 427-432.

[6] Zaccheria F., Mariani M., Psaro R., Bondioli P., Ravasio N., Environmentally friendly lubricants through a zero waste process, Appl. Catal. B., 2016, 181, 581–586

487349 Targeting Maximum Energy and Water Efficiencies for the Sustainable Total Textile Waste Refinery

Abstract:

The global textile industry is among the top ten manufacturing economic activities attaining a market value of €0.82 trillion. In 2008, 24 million tonnes of cellulosic and 39 million tonnes of synthetic fibres were sold globally (Oerlikon report, 2009). Despite industry's great economic value, the textile industry is among the most polluting ones. To only cultivate 1 kg of cotton; 16 g of pesticides, 460 g of fertilizer and 22 $m^3$of water is consumed (Saleem et al., 2010). In EU alone, 10 million tonnes of textile waste are produced each year (EASME, 2015). Specifically, in central Europe, Nordic countries, USA and Japan, the total post-consumer textile waste going to landfill is estimated 0.6, 1.9, 0.6, 0.1, 0.4, 0.9 million tonnes respectively. The present work coordinates established and original Process Integration technologies to the solution of a new problem; the design of the sustainable Total Textile Waste Refinery (TTWR); a new biorefinery type, that it is being currently developed in the course of the European Horizon2020 research project (RESYNTEX, 2015). It uses the complete spectrum of waste textiles as feedstock, aiming to provide a holistic response to the largely overlooked waste textile management problem. TTWR is now just before the phase of pilot construction and process integration technologies are currently implemented to accelerate the design solution.

TTWR consists of gate links between textile materials with commodity and specialty products and chemicals. Feedstock includes the entire range of textile fibres, consisting mostly of cotton, polyethylene terephthalate (PET), polyamides (PAs), wool but also other material in smaller quantities like acrylics and elastane. Glucose, amino-acid solutions, and PET and PA monomers and oligomers are the gate chemicals produced by the decomposition of the feedstock before the process eventually produces bio-ethanol, adhesives used in the wood-based panel production, PET bottles and PA-derived value added chemicals (caprolactam, amylamine, hexanoic acid etc.). The decomposition of natural and synthetic textile fibres is realised using patented chemistries with emphasis on chemical and enzymatic hydrolysis. The overall process consists of 20 sections dedicated to pretreatment for the reduction of the feedstock size, discolorations for the removal of dyes and impurities, depolymerizations for the decomposition of material and the production of gate chemicals, separations and purification of gate chemicals, reactions and separations for the production of end products and finally solid and liquid waste treatment processes. At particular stages (discolorations, separations and purifications), the TTWR process can be considered as the reverse process of textile production (pretreatment, dying, polishing etc.) and therefore, it has similar characteristics regarding water consumption. Without any water integration attempts the water consumption is as high as 52 tonnes per tonne of feedstock, and it is shared among reaction and separation stages (mostly washing and filtration steps) at a 2:1 ratio. Contrary to the textile fabrication processes which uses an average of 30 MJ/kg of fabric (Kalliala and Nousianinen, 1999), TTWR is more energy intensive, requiring more than 53 MJ/kg of feedstock. The majority of the energy requirements (~40%) is used in reactions, which are mainly neutral thermochemical hydrolyses at high temperatures for PAs and PET, and biochemical hydrolyses at mild temperatures. Considering that water is used at high ratios compared to the feedstock material in all reactions, it is also the main energy carrier in the reactors. This evidence indicates clear scope for a combined energy and water integration study, since water is used both as a cleaning agent for removing dyes and impurities as well as the main heating medium.

The TTWR process design procedure benefits from the rich technical know-how that has been accumulated so far for the design of conventional chemical processes. The present work takes advantage and improves on the significant contributions in process synthesis and integration (Mountraki et al., 2011a, 2011b; Tsakalova et al., 2012) made in the course of the European research project BIOCORE (Biocore, 2014) dedicated to the evolution of 2^{nd}generation biorefineries.

Novel process integration tools have been developed for assessing alternative process integration scenarios and achieving maximum efficiencies in energy and water use in modern biorefineries (Koufolioulios et al., 2014; Nikolakopoulos and Kokossis, 2016). Also, total waste management and recycle design technology has been produced to enable the development of designs for processes with minimum environmental impact (Mountraki et al., 2016; Nikolakopoulos et al., 2016a).

Water and energy Integration is applied with a combined use of process modelling targeting, process synthesis and mathematical optimization. Targeting sets the scope to match and improve efficiencies for energy and water (Pinch, Water Pinch), explaining efficiency bottlenecks, modifications required to improve targets, means to re-use and regenerate water and solvents, as well as means and technologies appropriate to treat waste.

A new tool is proposed In this paper for simultaneously targeting maximum energy and water efficiencies in the TTWR. It is developed by integrating a mathematical programming heat cascade sub-model for targeting and optimally allocating heating and cooling utilities (Koufolioulios et al., 2014) with a two-step iterative procedure to coordinate the targeting of minimum fresh water requirements, the calculation of recycle and treatment flows and the selection of water treatment processes (Nikolakopoulos et al., 2016b).

For the first model, combinations of utilities are built by matching pseudo-cooling utilities with steam levels so as to permit alternation of the purpose of the utility from cooling to heating and thus exploit excess of heat. This actually defines a dual function for a subset of cold utilities. The method calculates for each combination of utilities the minimum utility requirements through an expanded transhipment model of (Papoulias and Grossmann, 1983). The model incorporates the utilities as process streams with the objective of minimizing total utility cost. Heat receivers (i.e. cool streams matched against steam at the same level) are being maximized by being assigned a negative sign. Hot and cold utilities at the same level are assigned different costs and prices respectively. The model calculates the utility combination that minimizes the annualized total cost and maximizes profit.

The second model, consists of two interlinked transhipment models, where successively increasing recycle flows are introduced until convergence to the optimal recycle flow value based on a total cost criterion. The first is a transhipment model used to identify the flowrates and concentrations of the wastewater streams at the effluent water mains. The second model, originally intended for targeting treatment flowrates (Nikolakopoulos et al., 2014) is extended in screening treatment technologies and suggesting processes to integrate. The models demonstrated cascade characteristics by allowing the transfer of contaminant between concentrations intervals.

The application of the proposed tool in the design of the TTWR is able to produce significant energy and water saving targets, that may reach up to 50% for energy and 70% for water. An additional advantage of the procedure is the possibility to detect critical parameters of the system, where retrofitting changes improve on total cost. These are the number and order of washing and filtering stages and their concurrent or countercurrent mode of operation. The results of total water network suggest the use of centralized treatment using ultimately ultra and nano-filtration preceded by simple pretreatment filtering stages.

Acknowledgement

This project has received funding from the *European Union's Horizon 2020 research and innovation program*under grant agreement No 641942.

References

Biocore, 2014. EU research project FP7-241566, http://www.biocore-europe.org

Kalliala, E.M., Nousianinen, P., 1999. Life Cycle Assessment Environmental Profile of Cotton and Polyestercotton Fabrics. AUTREX Res. J. 1, 8–20.

Koufolioulios, D., Nikolakopoulos, A., Pyrgakis, K., Kokossis, A., 2014. A mathematical decomposition for the synthesis and the application of total site analysis on multi-product biorefineries, Computer Aided Chemical Engineering.

Mountraki, A., Tsakalova, M., Panteli, A., Papoutsi, A.I., Kokossis, A.C., 2016. Integrated Waste Management in Multiproduct Biorefineries: Systems Optimization and Analysis of a Real-Life Industrial Plant. Ind. Eng. Chem. Res. 55, 3478–3492.

Mountraki, A.D., Nikolakopoulos, A., Mlayah, B.B., Kokossis, A.C., 2011a. BIOCORE- A systems integration paradigm in the real-life development of a lignocellulosic biorefinery. Comput. Aided Chem. Eng. 29, 1381–1385.

Mountraki, A.D., Nikolakopoulos, A., Pyrgakis, K.A., Mlayah, B.B., Kokossis, A.C., 2011b. BIOCORE - On a paradigm for the design and synthesis of real-life biorefineries, in: 11AIChE - 2011 AIChE Annual Meeting, Conference Proceedings.

Nikolakopoulos, A., Faskiotis, D., Kokossis, A., 2016a. Integrated Transhipment Models for Synchronous Screening of Treatment Technologies and Targeting of Fresh Water and Recycle Flows, Computer Aided Chemical Engineering.

Nikolakopoulos, A., Faskiotis, D., Kokossis, A., 2016b. Cascade Models for Targeting and Synthesis of Total Water Networks. Chem. Eng. Trans. 52, 943–948.

Nikolakopoulos, A., Kokossis, A., 2016. A Problem Decomposition Approach for Developing Total Water Networks in Lignocellulosic Biorefineries. Process Saf. Environ. Prot. in press.

Nikolakopoulos, A., Thomaidis, L., Kokossis, A., 2014. Mathematical Programming Shortcut Screening Models for the Design of Integrated Waste Treatment Systems. Comput. Aided Chem. Eng., Computer Aided Chemical Engineering 34, 381–386.

Papoulias, S.A., Grossmann, I.E., 1983. A structural optimization approach in process synthesis—II. Comput. Chem. Eng. 7, 707–721.

Tsakalova, M., Nikolakopoulos, A., Kokossis, A., 2012. On the systematic synthesis screening and integration of real-life biorefineries, in: AIChE 2012 - 2012 AIChE Annual Meeting, Conference Proceedings.

495001 Advances in Process Intensification: Using Reactive Distillation for the Conversion of Crude Glycerol

Abstract:

Advances in Process Intensification: Using Reactive Distillation for the Conversion of Crude Glycerol

Obakore Agbroko and Tracy J Benson (Lamar University)

With the rise in biodiesel production over the last couple of decades has come the abundance of glycerol, a byproduct from biodiesel manufacturing. A typical biodiesel process includes the transesterification of triglycerides to produce fatty acid methyl esters, followed by fuel clean up steps: product separation, water washing, and acidulation. In short, the glycerol stream also includes water and methanol and is typically considered â€œcrude glycerol.â€ In this study, crude glycerol was converted to propanediols using isopropanol as a hydrogen donor molecule in a reactive distillation column.

Combining reaction engineering and distillation into a single unit operation presents some interesting challenges, yet, reduces the operational footprint as well as the operating and capital costs of manufacturing. Operational and design challenges include mismatched temperatures between the separating components and the reacting components, unfavorable volatilities, and slow reaction rates generally due to poor reactant/catalyst/product mass transfer rates. Reactive distillation is a likely pathway towards process intensification to minimize overall energy consumption, particularly since distillation accounts for 20 % of total energy used within the U.S. manufacturing sector. Increased reflux ratios typically yield higher product purities but at the cost of reboiler heat duties. In comparison with conventional approaches of reaction followed by distillation, the proposed scheme holds great potential due to its ability for lowering both operating and capital costs.

In this study, Aspen Plus was used to model the conversion of glycerol to 1,3-propranediol and 1,2-propanediol while being separated from methanol, water, and other inorganic components that are commonly found in waste glycerol from biodiesel production processes. Using the Non-Random Two Liquid activity coefficient model, the number of trays, conversion, and product selectivities were estimated. To verify simulation results, laboratory experiments were conducted using a 1-inch ID, packed distillation column. Column temperature profiles as well as reboiler and condenser heat duties were determined. The complete reactive distillation analysis will be presented as well as the extension of this technology into other areas of the chemical process industry.

495999 Process Design and Techno-Economic Assessment of a Seaweed-Based Biorefinery: Integration of Thermochemical and Biochemical Conversion Processes

Abstract:

Biofuel production using carbohydrate rich brown seaweed is being intensely researched in the last few years as a viable replacement for 1st and 2nd generation biomass. In this study, for the first time, integrated pyrolysis and anaerobic digestion is simulated on an industrial scale using Aspen Plus software and experimental data, and technoeconomical assessment is made. Main points of integration include the pyrolytic conversion of the digestate, instead of raw seaweed and further digestion of the carbohydrate rich aqueous phase, which is a byproduct of both pyrolysis and catalytic upgrading. The synergistic effect of this integrated process should yield a more economically and environmentally viable biofuel production, compared to standalone pathways.

491067 Process Intensification and Modularization for Sustainability

Abstract:

We will describe various concepts for intensifying catalytic reactions and reactors, including bifunctional materials and cascade reactions, fairly mature by now microtechnology, hybrids of reactors and separations, and multifunctional reactors including static micromixers and microheat exchangers (spatial domain), periodic operation (temporal domain), and use of alternate energy forms such as microwaves. These concepts will be elucidated with examples taken from biomass, shale gas, and ammonia chemistry as means of sustainable chemical processing.

489642 Assessing the Adequacy of a Sampling Frequency in the Application of PAT to a Continuous Manufacturing Process for Drug Product

Abstract:

The implementation of spectroscopic instruments to determine the concentration of active pharmaceutical ingredient (API) in powder samples is vast and well documented in the scientific literature. The introduction of continuous manufacturing lines for solid oral drug product has triggered increased interest in the application of such tools to determine the concentration of API in moving streams of powder. A key question in the development of such methods is the determination of an adequate frequency of sampling for the PAT devices, as it pertains the overall dynamics of the system

Our contribution will discuss a hybrid approach that uses information from a deterministic model, and the variability in the NIR spectra, along with time-series analysis techniques (such as ARX models, partial auto correlation coefficients, and cross-correlation coefficients) and the Shannon theorem of sampling to assess the adequacy of proposed frequencies of sampling and averaging in the spectrometer located in the feed-frame of the tablet press.

492244 Continuous Production Technology for Pharmaceutical Semi-Solid and Liquid Formulations: Processability and Influence of Process Parameters on the Product Quality

Abstract:

N. Bostijn[1,*], T. De Beer[1], W. Dhondt[2], J. Van Renterghem[1], C. Vervaet[3]

[1]Laboratory of Pharmaceutical Process Analytical Technology, Ghent University, Ottergemsesteenweg 460, 9000 Ghent, Belgium

[2]Janssen Global Technical Operations, Pharmaceutical Mfg (PM) Platform, Turnhoutseweg 30, 2340 Beerse, Belgium

[3]Laboratory of Pharmaceutical Technology, Ghent University, Ottergemsesteenweg 460, 9000 Ghent, Belgium

INTRODUCTION

Pharmaceutical production today mainly consists of batch processes. These are time-consuming, expensive and sub-optimal processes. Continuous processes on the other hand, allow for more flexibility, an easier scale-up and better quality control. The aim of this study is to investigate whether the Contimix system, a modular continuous production system, initially developed for the continuous production of semi-solid and liquid cosmetics, can be employed for pharmaceutical applications as well. A commercially available pharmaceutical cream, which is originally produced via a batch process, was manufactured with the Contimix system. In a first step, the processability of the cream with the Contimix system was investigated. Furthermore, the influence of the process parameters on the product quality was studied through statistical Design of Experiments (DoE), where the different process parameters were varied simultaneously.

MATERIALS AND METHODS

Materials

A pharmaceutical cream (Janssen Pharmaceutica N.V., Beerse, Belgium) was selected as model formulation.

Methods

All creams were produced with a continuous production system (Contimix, Hebold systems, Cuxhaven, Germany). A two level fractional factorial design was applied to study the influence of five process parameters on the product quality: process temperature, throughput, mixing speed in mixing vessel 1, mixing speed in mixing vessel 2 and water phase temperature. The studied quality attributes of the cream (DoE responses) were Active Pharmaceutical Ingredient (API) concentration (at the beginning, middle and end of a DoE run), yield point, viscosity of the yield point, viscosity (at low, intermediate and high shear rate) and pH. The data was analyzed using the MODDE Pro 11.0 software (MKS Umetrics, Umeå, Sweden). The API concentration of the creams was measured using a Raman Rxn2 spectrometer (Kaiser Optical Systems, Ann Arbor, MI, USA), equipped with a fiber-optic PhAT probe. The laser wavelength was 785 nm. Spectra were collected off-line with an exposure time of 20 s and a resolution of 5 cm^{-1}. The spectral range of the system was from 150-1890 cm^{-1}. Rheological measurements were conducted with a Haake Mars III rheometer (Thermo Fisher Scientific, Karlsruhe, Germany). The experiments were carried out at a fixed temperature (25°C), using a 60 mm parallel plate geometry. A stress ramp and viscosity curve were obtained from each produced cream. The pH of the creams was measured with a glass HI 3220 probe (Hanna instruments, Temse, Belgium).

RESULTS AND DISCUSSION

Prior to the experiments, the raw materials of the cream were divided into three pre-phases, which were batch mixed before the actual continuous production. This additional step was necessary since the Contimix system does not feature as many pumps as there are components in the cream. Furthermore, there is no possibility of introducing powders directly into the process. The results of the DoE show that the influence of the process parameters on the pH was negligible. The API concentration at the three time points between the experimental runs was not influenced by the process parameters. However, when the API concentration within the runs was evaluated, a downward trend in API concentration was detected for runs with a high throughput. This was caused by an inhomogeneity of pre-phase 3, which contained the API. Even for runs produced at a lower throughput, a similar trend was observed, suggesting that the homogenization of pre-phase 3 was not optimal, leading to an inhomogeneity of the API in the final product. The stress ramp experiments showed an influence of the process temperature. A higher process temperature was linked with a lower viscosity. The viscosity curve experiments showed an influence of the process temperature and throughput, but both influences were minimal. However, when the viscosity curves were investigated, a difference between the runs at low and high process temperature was visible. A possible explanation for the influence of the process temperature on the yield point and viscosity can be found in the short time between manufacturing the creams and conducting the rheological experiments. To draw conclusions about this effect, a number of runs at low and high process temperature should be executed and the viscosity of these creams should be measured directly (1 day) and a longer period (4 weeks) after production. Despite achieving processability, further improvements are necessary to transfer this continuous production technology to the pharmaceutical industry. The batch pre-step should be replaced by a continuous step. In order to achieve this, a way of feeding powder directly into the system is necessary. This would make the production system completely continuous, with all the advantages this entails. The original batch produced cream was measured with the same techniques as the creams of the DoE runs and compared with a center point run (run 18). The pH and API concentration were similar for both creams. From the stress ramp experiment, it can be concluded that the yield point of run 18 was slightly higher than that of the original product, indicating that both creams have similar rheological properties. This was confirmed by the viscosity curves, where the viscosity decrease with increasing shear rate was similar.

CONCLUSION

Processability of a pharmaceutical cream with the Contimix, a continuous production system, was achieved. From the screening design, it can be concluded that the process temperature had a significant influence on the viscosity of the creams, i.e. higher process temperature resulted in creams with a lower viscosity. Changing the other process parameters had no significant impact on the measured quality attributes. Furthermore, the Contimix was capable of producing a commercially available pharmaceutical cream in a continuous way with a similar quality as the original formulation.

494641 Modeling of Residence Time Distribution of a Continuous Dry Granulation Tableting Line

Abstract:

In recent time academia and industry have shown the economic and process related benefits of continuous pharmaceutical manufacturing. Regulatory authorities support strongly this approach of processing as well. This trend leads to a need for advanced control strategies, which enable automated operation of integrated production lines. With the first continuous plants being implemented for continuous production of solid oral dosage forms, out-of-spec material discharge strategies and process control with the aim to ensure robust product quality and to save resources have to be developed. Malfunctions of process equipment, unexpected process events and start-up or shut-down phases can lead to out-of-spec material. Due to material holdup and back mixing effects in specific unit operations, the out-of-spec material distribution can spread over process time. Therefore, knowledge of the residence time distribution (RTD) is essential not only for batch definition and material tracking aspects. For the development of cost-effective discharge strategies a fundamental tracking of the material through the process based on RTD is required.

In this work a dry granulation line, as one of the common ways of solid dosage form manufacturing, was investigated. The line consists of a feeding, blending, dry granulation, pneumatic transport and tableting step. For the determination of the residence time distribution in the considered unit operations as well as in the entire line a step change of API was performed at different total throughputs. As a system response the concentration of API was measured via in-line NIR spectroscopy at the outlet of blender, granulation and tablet press. Additionally, samples of powder, granules and tablets were drawn and analyzed off-line by means of UV/Vis as a reference.

Based on the experimental RTD data a transfer function was calculated and the transfer behavior of single unit operations modeled. The RTDs for single unit operations were interlinked and the entire process simulated with customized models in the process simulation software gSolids. The selected approach allows the prediction of out-of-spec material amount and the discharge time for different throughputs. This work shows promising results for both the inter- and extrapolation of RTDs for various throughputs and comparison with validation runs of the entire line.

494709 Robust State Estimation of Feeder and Blender Systems in Continuous Pharmaceutical Manufacturing Systems

Abstract:

Abstract

The development and application of advanced monitoring, operational- and design-making strategies for continuous pharmaceutical manufacturing require good understanding of the current state of the system. Moving horizon estimation (MHE) can estimate the current system state by solving a well-defined optimization problem where process complexities are explicitly taken into consideration as constraints. Traditional MHE techniques assume random measurement noise governed by some normal distributions. However, this assumption is not always valid since the actual distributions are often unknown or exhibiting departures from normal form. In addition, conventional state estimators fail to consider gross or systematic errors, such as bias, outliers, and drifts. The presence of gross errors can contaminate the state estimates and the reconciled measurements, causing undesirable â€œsmearingâ€ effects [1]. As a result, research on state estimation is closely related to topics of data reconciliation (DR) and gross error detection (GED).

Robust estimators, also known as the maximum likelihood estimators or M-estimators, are commonly used in reconciling measurements contaminated with gross errors. Robust estimators have been implemented in many research areas, including state estimation, data reconciliation, and parameter estimation [2, 3, 4]. Systematical comparisons between various robust estimators can be found in literature [1, 5]. To improve the accuracy and robustness of state estimation, we incorporate robust estimators within the standard MHE skeleton, leading to an extended MHE for dynamic state estimation with gross errors.

In continuous tablet or capsule production lines, the optimal operation of the feeder-blender system (FBS) is crucial in order to meet the critical quality attributes of the final product. Advanced monitoring and control strategies of FBS require fast and accurate estimates of the current process state. In this work, the proposed MHE framework is implemented on a simple FBS configuration which includes two loss-in-weight (LIW) feeders and one continuous blender. Based on our experiment results, we propose a first-principle dynamic model for screw feeders and a reduced-order, compartment-based model for continuous blenders. Both dynamic models are used in our MHE framework, to result in a dynamic optimization (DO) problem constrained by differential algebraic equations (DAEs).

The proposed MHE strategy is first tested on simulation runs of the FBS where measurements are contaminated with gross errors such as outliers and drifts. Numerical results show significant benefits in applying the extended MHE technique with robust estimators. In addition, we implement our MHE framework in a pilot plant FBS, which is part of a continuous dry granulation line, and demonstrate on-line dynamic state estimation and data reconciliation. Experimental results confirm that the proposed MHE framework is robust to gross errors and can provide accuracy state estimates. Moreover, the efficient solution of the MHE realized in this work, suggests feasible application of on-line state estimation on more complex continuous pharmaceutical processes.

Bibliography

[1] Özyurt, Derya B., and Ralph W. Pike. "Theory and practice of simultaneous data reconciliation and gross error detection for chemical processes." Computers & chemical engineering 28.3 (2004): 381-402.

[2] Tjoa, I. B., and L. T. Biegler. "Simultaneous strategies for data reconciliation and gross error detection of nonlinear systems." Computers & chemical engineering 15.10 (1991): 679-690.

[3] Arora, Nikhil, and Lorenz T. Biegler. "Redescending estimators for data reconciliation and parameter estimation." Computers & Chemical Engineering 25.11 (2001): 1585-1599.

[4] Nicholson, Bethany, Rodrigo López-Negrete, and Lorenz T. Biegler. "On-line state estimation of nonlinear dynamic systems with gross errors." Computers & Chemical Engineering 70 (2014): 149-159.

[5] Prata, Diego Martinez, José Carlos Pinto, and Enrique Luis Lima. "Comparative analysis of robust estimators on nonlinear dynamic data reconciliation." Computer Aided Chemical Engineering 25 (2008): 501-506.

496016 Large-Scale Experimental Comparison of Batch and Continuous Technologies for Pharmaceutical Tablet Manufacturing

Abstract:

This work presents experimental comparison of batch and continuous technologies for pharmaceutical tablet manufacturing at a scale of 100 kg. The investigation covered the units of mixing, granulation, drying, blending, and compression, for which three processes were established based on batch and continuous technologies. For the batch process, conventional facilities were used, with employing fluidized bed granulation and high shear granulation as the granulation unit. For the continuous process, a recently developed facility was used, which interconnects all the units and is capable of producing tablets at the speed of 25 kg/hr. Common to these three processes, a 100 kg of raw materials were provided that contained ethenzamide as an active pharmaceutical ingredient (API), and mannitol and crystalline cellulose as excipients. As a preparation, experiments at a small scale were performed in order to determine process parameters in the experiments at a scale of 100 kg. During or after the production, samples were taken for investigating the following three items. First was the quality of granules as intermediate products, e.g., particle size distribution, mean diameter, and loose bulk density. Second was the quality of the tablets, e.g., tablet hardness, content of API, and dissolution profiles. Third was the yield of the overall process considering various causes of losses, such as stabilization of the continuous process, sticking to the equipment, or condition setting of compression. For the continuous process, stability of product quality during the process operation was also investigated because the total run time can be up to hours.

The experimental result indicated that the product quality of all processes fulfilled the standard values which are often used in the industry. Granules made by fluidized bed granulation were porous, while granules made by high shear granulation and the continuous process were spherical. Although hardness of tablets made by the high shear granulation process was lower than that by the other two processes, there are no significant differences of dissolution profiles among the three processes. In the continuous process, product quality measured by mean particle diameter of granules was stable for the entire run time except for the initial nine minutes. Because of this reason, yield of the continuous process was lower than those of the other two processes, which indicates that stabilization would be one of the important challenges in continuous technology. In conclusion, the three investigated processes using continuous, fluidized bed, and high sheer granulation showed similarity in terms of product quality. However, intermediate products showed different quality characteristics such as size/shape, and especially in continuous technology, yield was found to be an improvement opportunity. The actual decisions should be based on various aspects such as quality, economy or safety, and thus, development of multiobjective design methodology is desired in future.

496075 A Cost-Benefit Analysis of Continuous Manufacturing Using Discrete Event Simulation

Abstract:

In the pharmaceutical industry, the demand for small volume, market specific products has increased in recent years. However, current batch processes do not allow the flexibility to meet smaller demands without a high cost of inventory and risk of discards. Continuous processing may offer advantages to traditional batch processing including shorter lead times, reductions in inventory, decreases in management overhead, and fewer discards or loss of tenders. However, the specific conditions which make continuous manufacturing beneficial are currently not well understood. To quantify the benefit and determine the conditions where continuous manufacturing is an attractive alternative for products, a discrete event simulation model was developed. The model compares continuous to batch tablet formulation processes under numerous different scenarios including; various market conditions, production planning strategies, capacity constraints, and production volumes. The model uses a product currently produced at a Merck production facility as a base case. The outputs from the model define the product, market, and production conditions best suited for a continuous manufacturing platform. This ultimately enables a mechanism for identifying products that are a good fit for a continuous manufacturing platform.

496960 Real-Time Monitoring and Control of API Concentration in a Tablet Press for Continuous Manufacturing of Tablets

Abstract:

Active pharmaceutical ingredient (API) concentration in powder blends is a critical material attribute (CMA) that is considered one of the most important product critical quality attributes (CQAs) for their final product (tablets). To determine the API concentration before tableting, real-time monitoring of API concentration before and after compaction has been performed for a variety of batch and continuous processes via Near-Infrared (NIR) spectroscopy [1, 2]. This analytical technology, which has been highly encouraged within the Quality by Design (QbD) and Process Analytical Technology (PAT) frameworks, has shown to be a valuable for detecting issues in CMAs during the process and has become a critical in-process sensor in the pharmaceutical industry [3].

This work presents a feed-forward controller for continuous manufacturing by using a NIR-based real-time monitoring technology at a tablet press feed frame. The basic idea behind feed forward control is to measure important process variables and take corrective action before the disturbances upset the process. The feed forward controller therefore needs to receive information about the known and predictable effects of process disturbances before those occur [4]. The proposed methodology involves measuring the inherent variability in the API concentration of material coming from the process and adjusting tablet weight to maintain overall API potency in tablets. We measured the residence time distribution (RTD) information from the rotary tablet press together with the NIR from the feed frame to develop the proposed framework as follows.

We evaluated the RTD of our Fette 1200i tablet press unit. A step change of API concentration of powder blend was given before the feed frame and the API concentration by the feed-frame NIR and compressed tablets were measured [5]. Based on these profiles, RTD of the feed frame was shown to have a similar RTD profile as a continuous stirred tank reactor (CSTR). RTD of the turret was found to be negligibly short and no difference was observed between the profiles of the feed-frame and tablets. The estimated RTDs are useful for prediction of the CQA and its controls. The formulation for our case study consisted of micronized acetaminophen, microcrystalline cellulose (Avicel® PH102), and magnesium stearate at a ratio of 15:84:1 by weight, respectively. A portable JDSU NIR probe was attached in the feed frame, sampling the powder spectra through one of the unit's glass windows. Five (5) different blends ranging from 10 – 20% API concentration were made as calibrations for the NIR model. NIR spectra were measured in-line using the 5 powders blends to build the calibration model *in situ*. A calibration model with small calibration error was constructed by partial least squares (PLS) analysis. Validation experiments with four different acetaminophen concentrations demonstrated that the constructed model has high accuracy during long-term operation.

Overall, based on the feed-frame's RTD profile and the information from the NIR sensor's model, a feed-forward controller was developed to account for changes in powder blend API concentration and adjust tablet weight to decrease variability of API potency in tablets. It was verified by both software simulations and experiments that API potency could be maintained at the target value by the feed-forward control compared to the open-loop process.

References

1. Vanarase, A.U., et al., *Real-time monitoring of drug concentration in a continuous powder mixing process using NIR spectroscopy.* Chemical Engineering Science, 2010. **65**(21): p. 5728-5733.

2. Jarvinen, K., et al., *In-line monitoring of the drug content of powder mixtures and tablets by near-infrared spectroscopy during the continuous direct compression tableting process.* Eur J Pharm Sci, 2013. **48**(4-5): p. 680-8.

3. USFDA, *Guidance for Industry: PAT — A Framework for Innovative Pharmaceutical Development, Manufacturing, and Quality Assurance*, U.D.o.H.a.H. Services, Editor. 2004: Rockville, MD.

4. Myerson, A.S., et al., *Control Systems Engineering in Continuous Pharmaceutical Manufacturing. J. Pharm. Sci.*, 2015. **104**: p. 832-839.

5. Ward, H.W., et al., *Monitoring blend potency in a tablet press feed frame using near infrared spectroscopy.* Journal of Pharmaceutical and Biomedical Analysis, 2013. **80**: p. 18-23.

Pacific Northwest NATIONAL LABORATORY
Proudly Operated by Battelle Since 1965

Modularized Production of Value-Added Products and Fuels from Distributed, Waste Carbon-rich feedstocks

Robert Weber[1], Johnathan Holladay[1], Cynthia Jenks[2]

1) Pacific Northwest National Laboratory; 2) Argonne National Laboratory

Topical Conference Session: Distributed Chemical and Energy Processes for Sustainability, AIChE National Meeting, October, 2017

Overview

Valorizing waste requires distributed facilities that, in turn, require new catalysts and new reactors to achieve economic, robust operation.

- We seek to locally convert waste carbon into chemical and fuel precursors, which aggregate to about 25% of US demand.
- Throughput will come from numbering up, not scaling up, which should accelerate deployment by attenuating technical and investment risk.
- The remoteness, dispersion and heterogeneity of the resources requires development of new catalysts, reactors and intensified processes.
- For example, electrocatalytic upgrading of liquefaction bio-oil could integrate upgrading with clean-up, further "preventing a waste and turning it to profit".
- This project probes both fundamental science and the practicality of electro-upgrading of the components of bio-oil *ex* hydrothermal liquefaction.

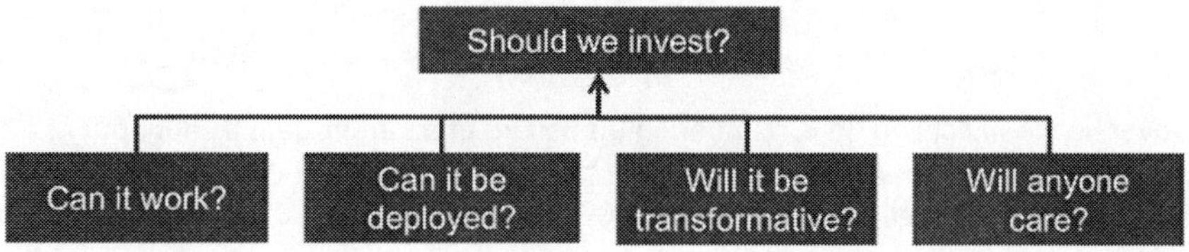

1/17

Will anyone care: *size of the opportunity*

We seek to locally convert waste carbon resources into chemical and fuel precursors, which sum to about 30% of US demand.

Type	Source	Availability (Mboe/y)	Site capacity (boe/d)	Possible Process	Tech. issues
Dry	MSW, ag & forest residue	1250	2-20	Pyrolyze, Hydrotreat	Hetero-geneity
Wet	Food, sludge manure	750	2-150	HTL, Hydrotreat	Product slate
Gas	Stranded CH_4 biogas, CO_2	50	70-1500 (CH_4) 1-10 (CO_2)	Reforming FT, Magic	Product slate
Total		**2000**			
US annual petroleum import		6900	130,000		
US annual net import of oil		1700			
Fuel demand of US cars		2800			
Energy demand of US homes		2000			

2/17

Will it be transformative: *benefits of modularity*

To attenuate technical and investment risk and to benefit from manufacturing efficiencies and linear scaling, we will number up rather than scale up.

- Replicating many (≳1000) small facilities promises more attractive investment and risk profiles than are typically found for scaling up for small scale.

Nominal risk profiles: most the risk is resolved in the first deployment for modular systems.

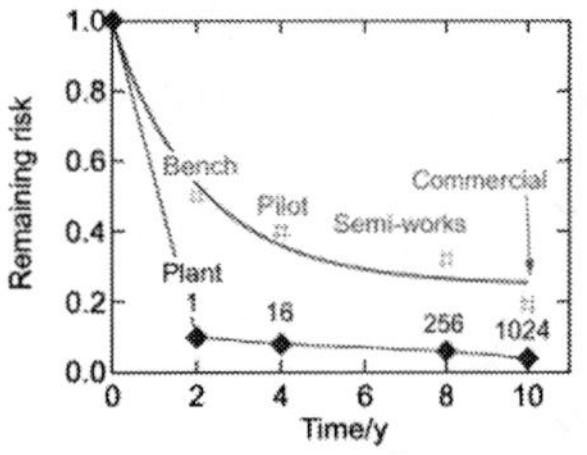

Nominal cost profiles for modular and stick-built plants each summing to 2000 bbl/day

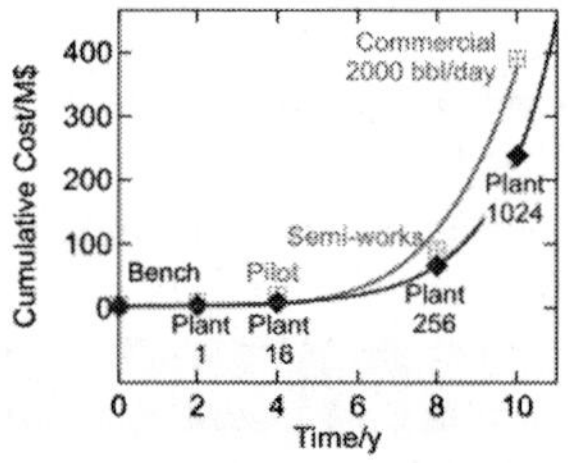

Having a large number of modular plants would benefit from economies of manufacture

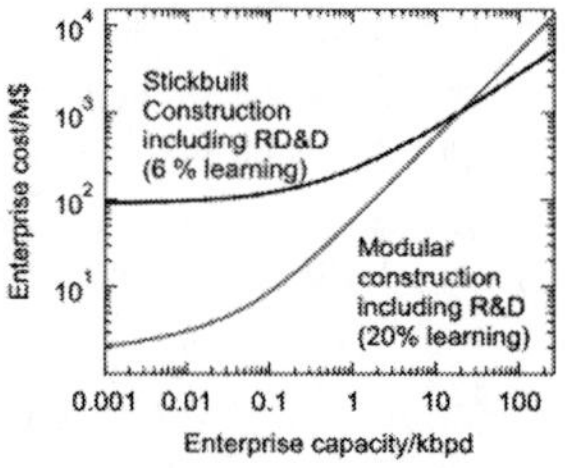

- The cross-over between scaling up & numbering up occurs at an aggregate throughput of >20 kbbl/day.
- Going from 2000 bbl/day to an aggregated 4000 bbl/day would cost ~$460 million to add 1000 more 2 bbl/day facilities; a second commercial plant (at a 6% learning rate) would cost “only” $280 million.

3/17

Can it work: *science and technology needs*

The remoteness, dispersion and heterogeneity of the resources requires development of new catalysts, reactors and intensified processes.

- The size of the facility will be dictated by the supply rate of feedstock, which will likely be neither homogeneous nor constant.
- Remoteness means we will not have a steam or H_2 grid or ways to valorize waste heat.
- The goal is to convert stranded carbon to fuel so we want to be as frugal as possible and not generate additional, even more dilute waste streams.

Challenge or technical need	Example of basic research
Separation processes and controls to facilitate conversion processes	Remote, online quantification of types and flow rates of solids and suspensions
Catalytic conversions with high efficiency, rates and selectivity under near ambient conditions	**Research on low temperature catalysis and chain reactions in liquid media, particularly in water**
Separation processes facilitating product valorization	Thermodynamic of and transport in complex mixtures
Water recovery and cleanup, again with high energy efficiency	Membrane science for separation of water from both polar and non-polar organics

4/17

Can it work: *implementing modularity*

To achieve the envisaged modularity requires changing from volume-based rates to area-based processes which downsize economically.

Modular facilities will incorporate Intensified unit operations, amenable to mass manufacture and which are cheaper than volume-based technologies at small scale

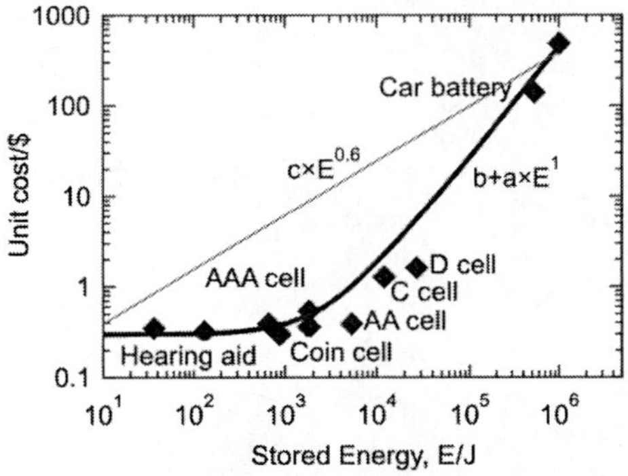

A similar scaling rule is observed for process heat exchangers (including consumer refrigerators) and other electrochemical reactors (fuel cells, flow batteries).

5/17

Can it work: *System architecture*

A complete system would include feed preprocessing, processing, upgrading and separations.

- Sites will be geographically distributed, likely without access to industrial utilities except electricity.
- The scale will be small (say, the equivalent to ~10 bbl oil/day).
- Operators may not be highly trained or closely supervised.
- The customers will be frugal.

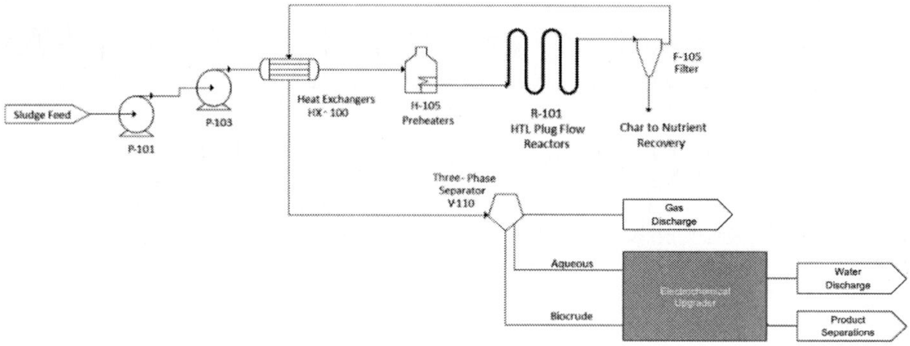

6/17

Can it work: *Scope*

We are now researching the downstream, upgrading and separations units to complement prior work at PNNL on the upstream conversion operations.

PNNL's Modular Hydrothermal Liquefaction

3 skids

- Feed prep and handling
- Hydrothermal liquefaction
- Biocrude recovery

Capability

- ~16 liters/h ≈ 2.4 bbl/day
- Solid-liquid separations at operating conditions
- Liquid-liquid separations for biocrudes that are heavier, or lighter, than water
- Product-feed heat exchange

7/17

Can it work: *electrolysis reactors*

We have constructed 3 scales of flow reactors that can provide both fundamental information and grow to practicable levels of capacity.

- Ag/AgCl reference electrode that can be positioned axially
- Carbon felt- or Carbon cloth-supported catalysts
- Automated data acquisition and control over E, q and T.

Bench: 2.4 cm × 4.2 cm

Pilot:, 5.7 cm × 10 cm

Demo: 10.7 cm × 18.7 cm

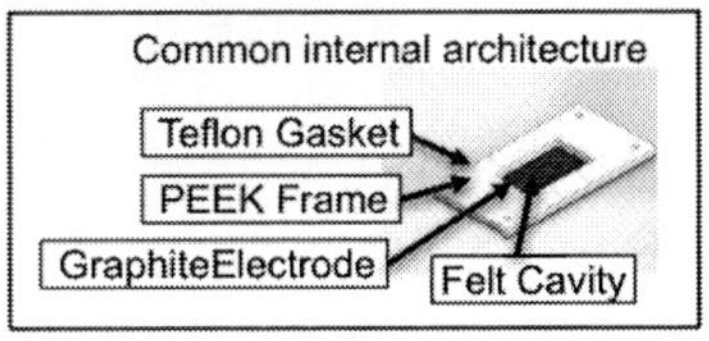

Reference electrode

8/17

Can it work: *electrolysis reactors*

We are in the process of characterizing, modeling & testing catalysts and reactors.

- The catalysts consist of redox sites and electrically conductive supports, which stabilize the redox sites and construct the solution environment.
  - E.g., metals and alloys deposited on nanoporous and fuzzy carbons; akin to PEMFC catalysts but with *high* overpotentials for making H_2
- Characterization will entail site-counting as well as site description, in both cases at least *in situ* and better *in operando*, e.g., Raman, SFG, XAS, coulometry and (with care) electron microscopy
- Modeling needs to estimate ΔG, not just ΔE, of reactions in fields $\geq 10^8$ V/m
- Testing will require reactors that afford intrinsic kinetics and that are scalable.

Visualization of flow in a lab-scale electrolysis reactor using Rhodamine 6G

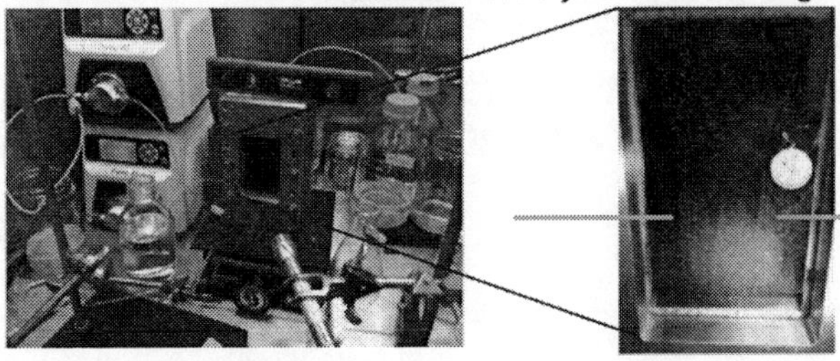

Nearly flat front is consistent with plug flow

9/17

Can it work: *Kinetics*

The rates of the model reactions correlate well with Butler-Volmer kinetics.

The reactors appear to be nearly plug flow; when the reaction rate laws accord with pseudo-first order kinetics, conversion, $C_R/C_i = \exp(-k\tau)$, where the rate constant, k, derives from Butler-Volmer kinetics

$$C_R = \exp\left[-k_0\left(\left(\exp\left(\frac{-\alpha F\eta}{RT}\right)+\exp\left(\frac{(1-\alpha)F\eta}{RT}\right)\right)+C_i\exp\left(\frac{-\alpha F\eta}{RT}\right)\right)\tau\right]$$

Kinetics fit using the bench cell (V =5 mL)
$0.05 < C_0/M < 0.253$; $0.1 < E/V < 1.1$;
$150 < \tau/s < 450$; $0.01 <$ loading < 0.04

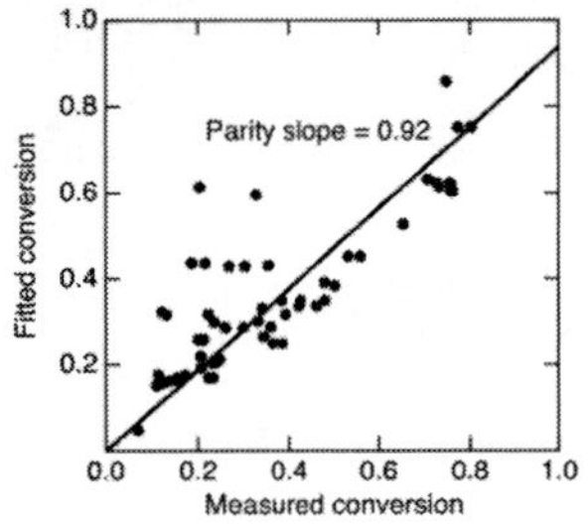

Comparison of the bench cell (5 mL) with the pilot cell (27 mL), each containing 4 wt% Cu/felt

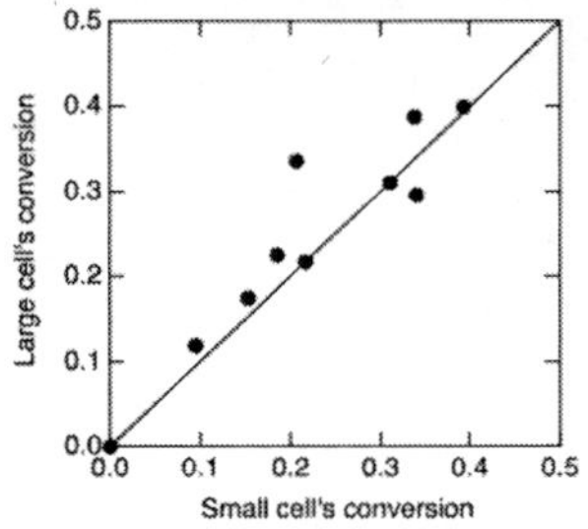

Can it work: *Cathode processes*

We have recently shown that we can electrochemically upgrade some of the functional groups present in neat bio-oil.

- Catalyst was 4 wt% Pd/C, 150 mL of whole bio-oil (ex pyrolysis of pine) recirculated for 20 h through a 6 mL cell at 1.5 V.
- The residence time $\tau = 20\ \text{h} \times 6\ \text{mL}/156\ \text{mL} = 0.8\ \text{h}$ (*cf* 5 h for thermal hydrogenation).
- ^{13}C NMR and ^{1}H NMR indicate a significant decrease in carbonyl and aromatic species post electrolysis.

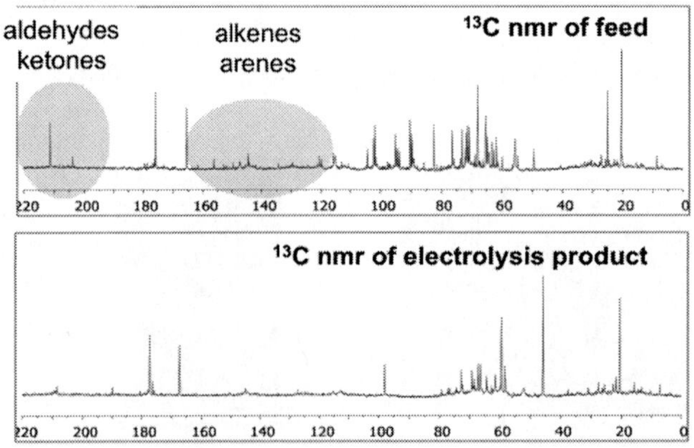

Can it work: *Anode processes*

The aqueous phase from liquefaction of wastes contains many compounds that can, in principle, be oxidized on the anode …

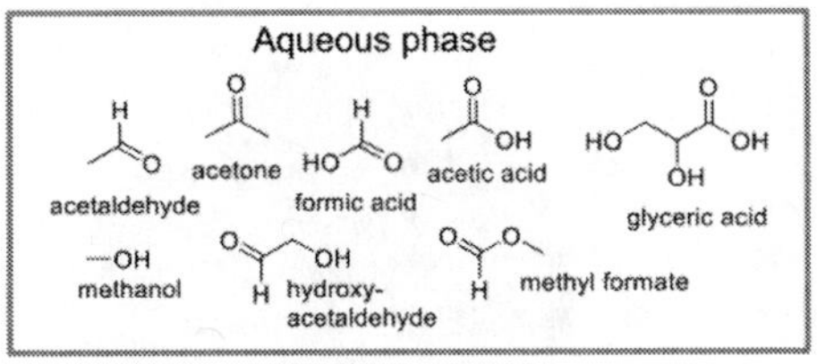

Stoichiometry of upgrading HTL bio-oil *ex* WWTP sludge (0.4 kg oil/kg sludge)

Stream	H^+ Equivalents/g_{sludge}
Oil	-0.03
Aqueous	+24

Estimated from Snowden-Swan et al., PNNL-25464 **2016**.

Can it be deployed: *process integration*

Electrocatalytic upgrading of the bio-oil could simultaneously remediate the wastewater further "preventing a waste and turning it to profit".[1]

- Electrochemistry provides two types of process intensification: high energy flux at ambient conditions (1 eV = 100 kJ/mol) and integrated operations.

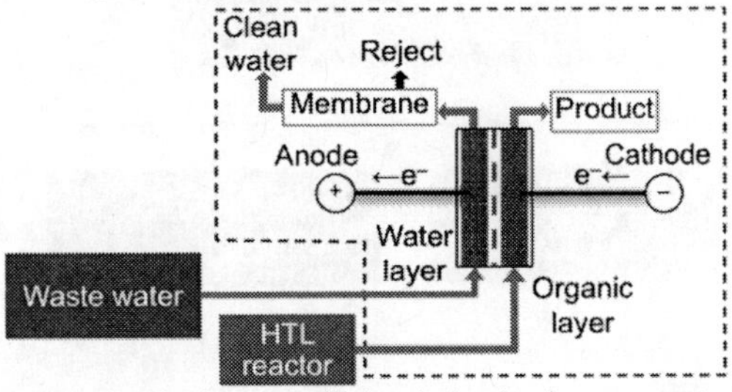

- The cathode catalyst must selectively and efficiently hydrodeoxygenate and hydrogenate the feed.
- The anode catalyst must preferentially and efficiently oxidize the dissolved polar compounds.
- The electrolysis reactor must be compact, robust and ensure good contacting between the liquid phases and their respective catalysts.

1) Arthur D. Little, *J Ind Eng Chem*, **1909**, *1*, 61-62."Every waste that is prevented or turned to profit, … and every more effective process which is developed, makes for better living in the material sense and for more wholesome living in the higher sense."

Can it be deployed: *Techno-economics*

Experience with redox flow batteries indicate that an electrochemical upgrader would cost about 1/3rd of the target CapEx = $50,000/bpd.

- Values for the electrochemical process were based on PNNL's analysis of the hardware costs for a redox flow battery.
  - Raising the heating value of 1 T/day of pyrolysis oil (~15 MJ/kg) to that of a fuel (~45 MJ/kg) requires adding 30 GJ in 86400 s = 350 kW
  - Estimated, near term CapEx for a redox flow battery, less electrolyte, = ~300 $/kW (exclusive of any PGM catalysts)
- An electrochemical upgrader based on similar technology would cost ~105 k$/T/day = $15,000/bpd.

CapEx estimate for a Redox Flow battery

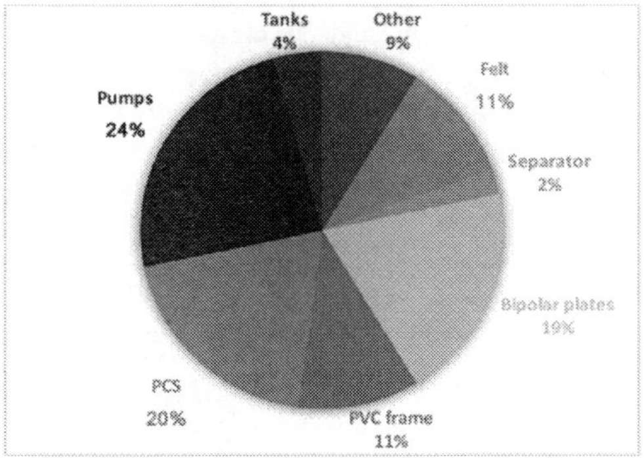

After V. Viswanathan et al., *J. Power Sources* **2014**, *247*, 1040

14/17

Summary

Downsized technology based on hydrothermal liquefaction and electrochemical upgrading appears to be a plausible way to valorize waste.

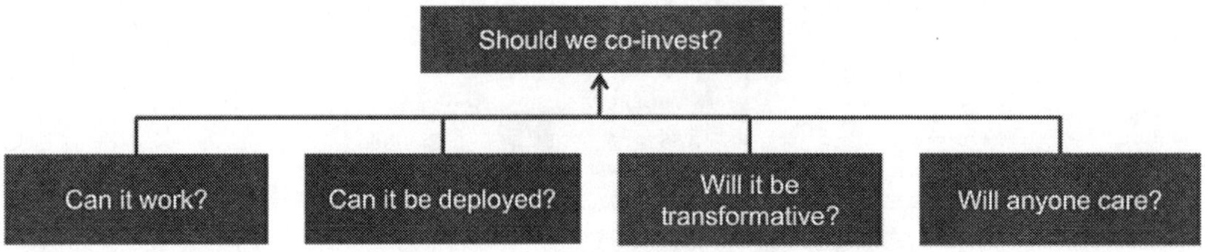

- We do need to devise new catalysts
- Reactors exist that could deploy them
- Such a new enterprise would be transformative: specifically it could convert a waste that now costs $60/ton to dispose and turn it into a profit stream.
- The new enterprise is being designed to appeal to waste-producing companies (e.g., agriculture), communities and product off-takers.

15/17

Acknowledgements

The research was supported by the Chemical Transformations Initiative at PNNL and was conducted under the Laboratory Directed Research and Development Program

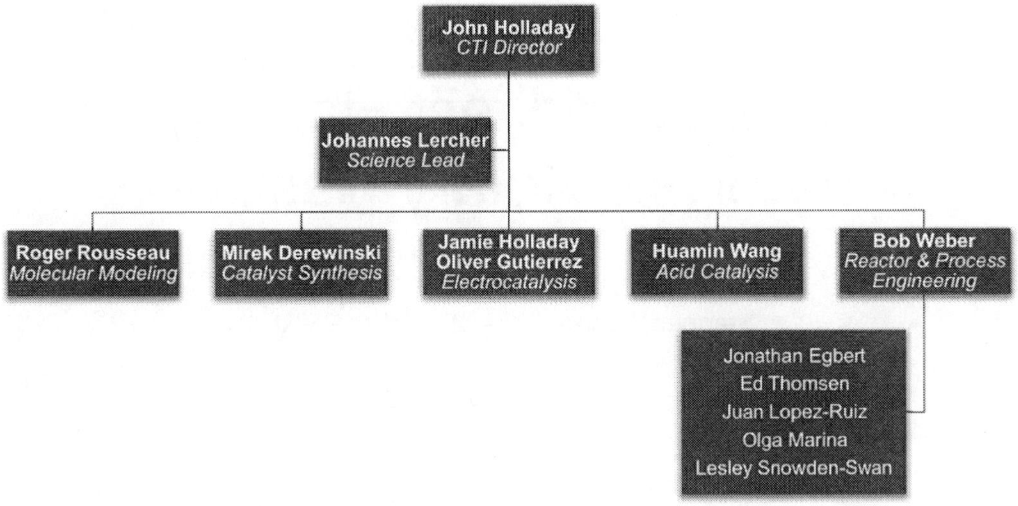

16/17

PNNL Overview

Pacific Northwest National Laboratory has been managed by Battelle for 50 years.

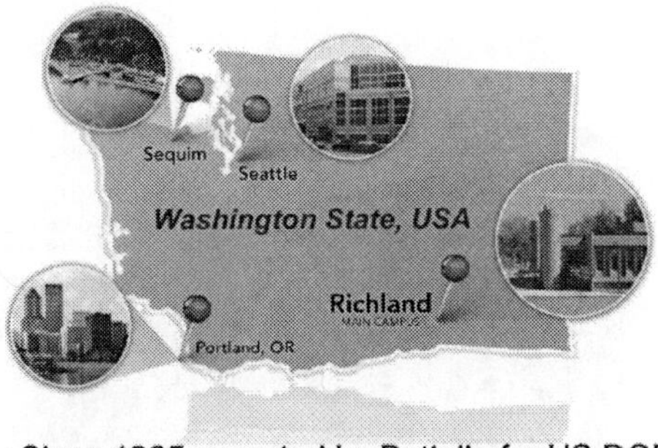

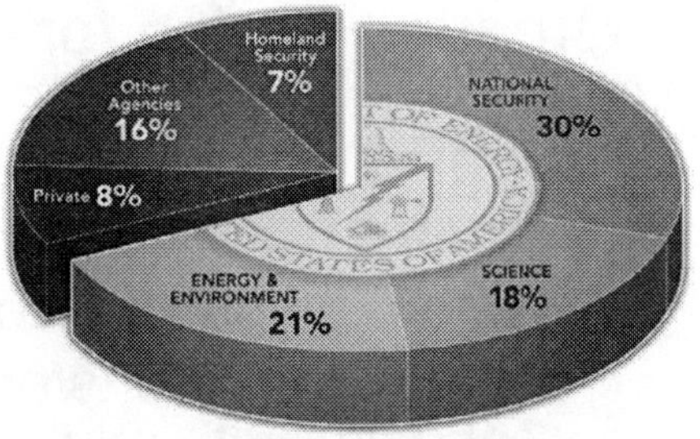

- Since 1965 operated by Battelle for US DOE
- Mission-driven collaborations with government, industry and academia
- Unique Science & Technology strengths and capabilities
- Operating budget of ~$1 B supporting a staff of 4,000
- ~900 peer-reviewed publications per year
- ~100 U.S. and foreign patents per year (2,400 since 1965)
- ~100 awards (R&D 100 and FLC)

17/17

Inclined Indirect Flaming Pyrolysis Rotary Gasifier

SERDP Project #WP-2211
Stephen Cosper, USACE-ERDC-CERL
Dr. Paul Amodeo, SUNY Cobleskill

Objective

- Create a design for a deployable waste-to-energy-system, with the following criteria:
  - Based on gasification
  - One to three tons/day capacity, with minimal sorting
  - Self-powered, with net-positive exportable electricity
  - Size to fit two, 20-ft containers

2

Technical Approach

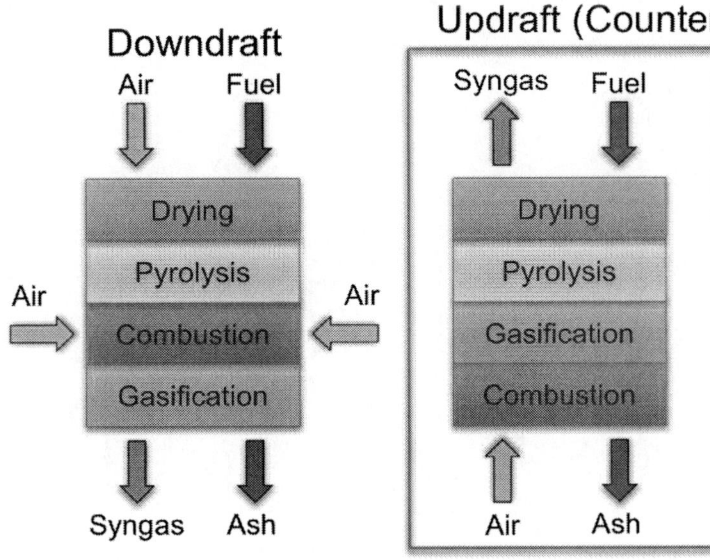

Countercurrent gasification has a number of inherent characteristics that would be beneficial for WTE systems:

- High thermal efficiency
- Less sensitive to fuel pretreatment
- Reduced particulate entrainment
- Low-reactivity ash produced from final-stage combustion

(Courtesy: UND-EERC) 3

Technical Approach – Rotary System

- System works like a modified updraft gasifier
- Waste fed from bottom, placed at hot, burning char layer
- Syngas collected through concentric tube, not directly exposed to char bed
- Rotates at ~1rpm

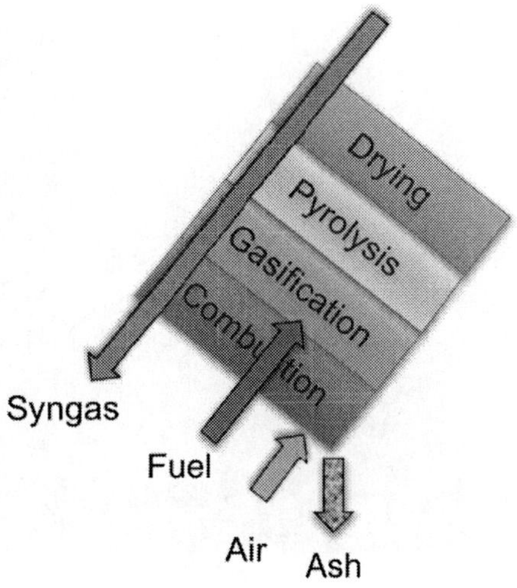

4

SERDP DOD • EPA • DOE | ESTCP

Initial design

- Initial hardware design and layout
- Experimental layout on two trailers to mimic portability requirements
- 1st trailer has hydraulic waste compression, gasifier, gas treatment
- 2nd has electronic control system, 60kW genset, hydraulic pump and controls, experimental electric load center

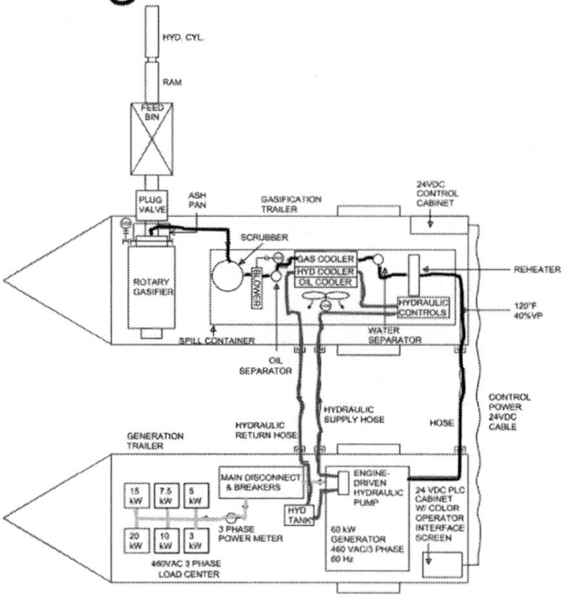

5

SERDP DOD • EPA • DOE | ESTCP

Waste Feed

6

SERDP DOD • EPA • DOE | ESTCP

Main Reactor

7

SERDP DOD • EPA • DOE | ESTCP

"Standard" waste composition

- Based on field surveys, 2013
- Variations included
    - 50% food
    - 50% plastics
    - 50% used oil

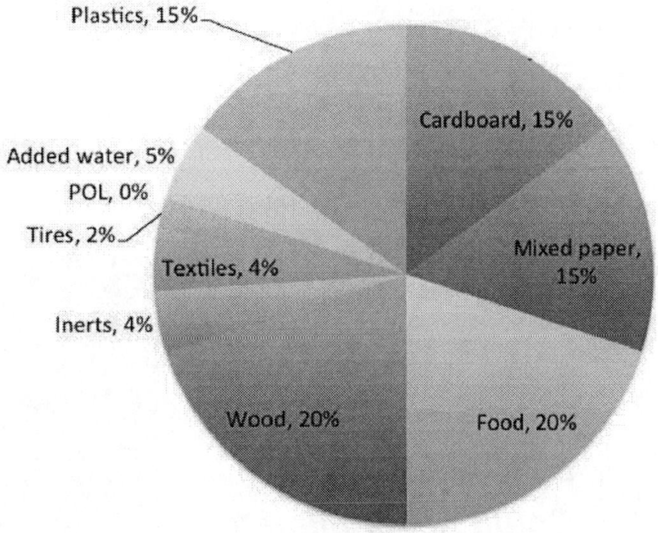

8

SERDP DOD • EPA • DOE | ESTCP

Temperature monitoring reactor

- Thermocouples down the length of reactor
- Steady state gas discharge temperature about 400°F
- Temperature peaks indicate time to add waste.
- High temps at end of run indicate, allowing waste layer to burn down.

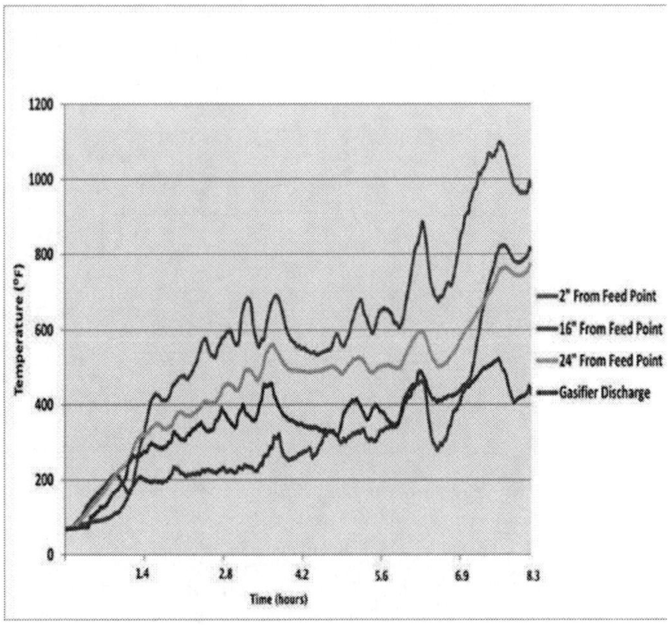

9

SERDP DOD • EPA • DOE | ESTCP

Quencher Tank / Pyro Oil Recycling

- Pyrolysis oils produced in reactor
- Accumulate in quencher tank
- Excess fed back into reactor
  - Eliminates a waste stream
  - Enriches syngas
  - ***Could be distilled to commercial fuels***

10

Quencher oil temperature

- Incoming syngas ~400°F
- Pyro-oil remains constant due to water emulsion
- Temp remains below boiling point of water

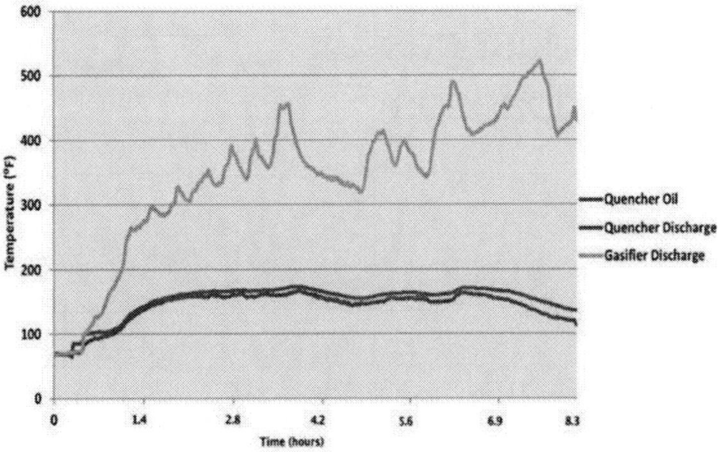

11

Comparison to Commercial Fuels

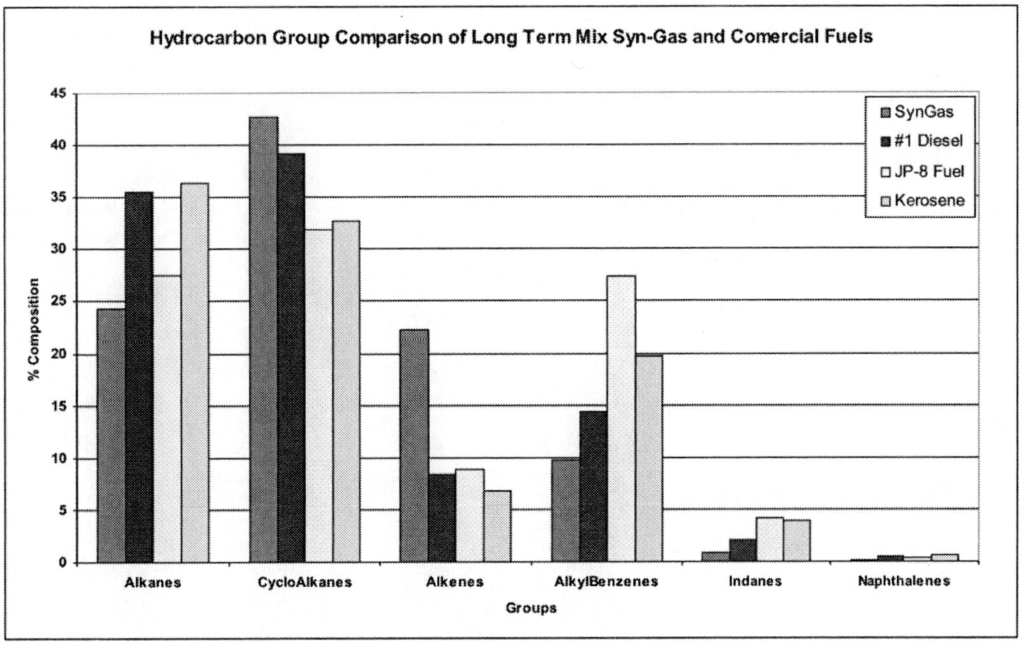

12

Exhaust gas comparison

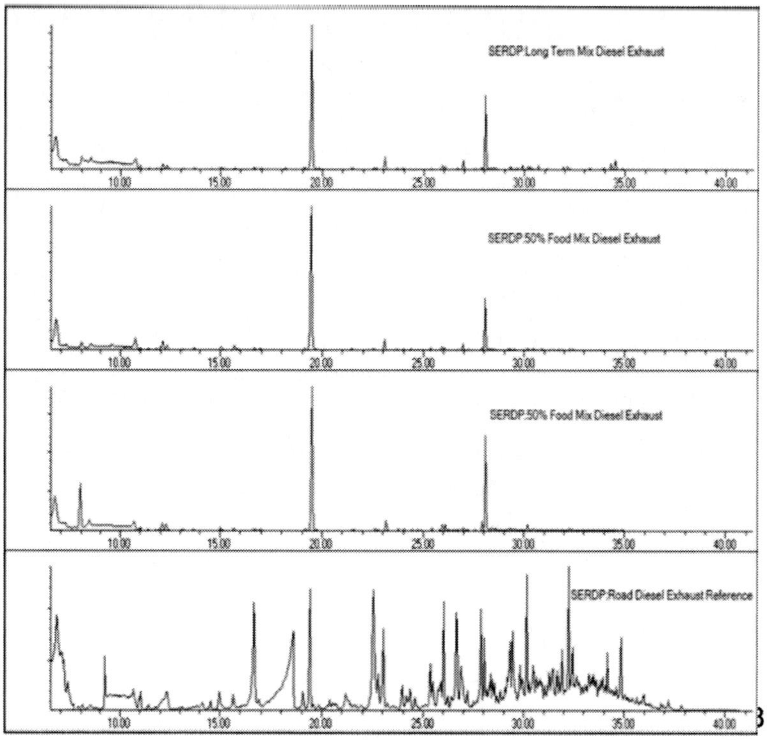

3

NO_X Emissions with Dual-fuel

- About 50% reduction in NO_X
- Due to:
  - less N_2 into engine
  - Cooler combustion temperatures

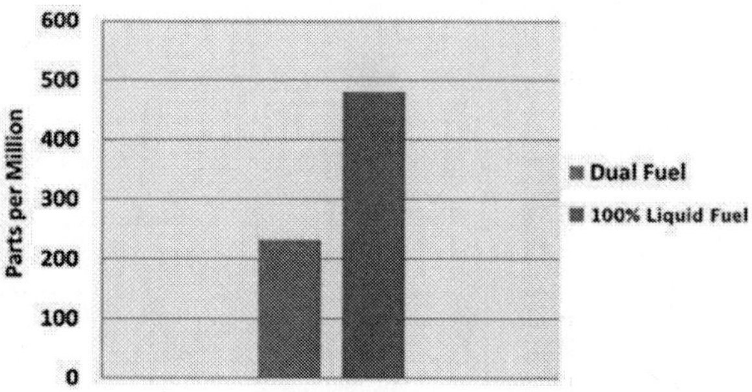

14

Status?

- Project complete
- Prototype designed, built, and tested on a variety of materials
- Final report published very soon
- Team is actively seeking funds for demonstration and/or to construct a "Mark II" automated, containerized version

15

Contact

Stephen Cosper
Army Corps of Engineers
Construction Engineering Research Lab
Champaign, IL
Stephen.D.Cosper@erdc.dren.mil
217-398-5569 M-Th
217-300-3721 F

Dr. Paul Amodeo
SUNY Cobleskill
518-255-5384
amodeopa@cobleskill.edu

https://www.serdp-estcp.org/Program-Areas/Weapons-Systems-and-Platforms/Waste-Reduction-and-Treatment-in-DoD-Operations/WP-2211/WP-2211

16

Monolithic Catalysts Coated with Hierarchical ZSM-5 for Fischer-Tropsch Synthesis

Chunxiang Zhu, David Gamliel, Julia A. Valla and George M. Bollas

Department of Chemical & Biomolecular Engineering
University of Connecticut

http://pdsol.engr.uconn.edu

10/31/2017
2017 AIChE Annual Meeting

UCONN

πδσωλ

Motivation

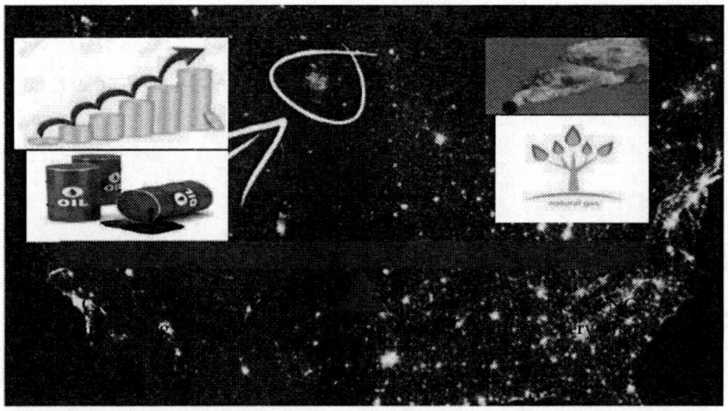

http://www.npr.org/sections/krulwich/2013/01/16/169511949/a-mysterious-patch-of-light-shows-up-in-the-north-dakota-dark
https://energyinfrapost.com/india-comfortable-crude-oil-prices-65barrel-cea/: oil barrel
https://www.123rf.com/photo_12683470_stacks-of-gold-coins-bullion-with-red-arrow-depicting-price-value-going-up-illustration.html : price goes up
https://www.google.com/search: nature gas

10/31/2017
2017 AIChE Annual Meeting

UCONN

πδσωλ

FTS Reaction Mechanism

Reaction	Stoichoimetry	Desirable?
Alkane Formation	$nCO + (2n+1)H_2 \rightarrow C_nH_{2n+2} + nH_2O$	YES
Alkene Formation	$nCO + 2nH_2 \rightarrow C_nH_{2n} + nH_2O$	YES
Water Gas Shift	$CO + H_2O \rightarrow CO_2 + H_2$	NO
Boudouard Reaction	$2CO \rightarrow CO_2 + C$	NO

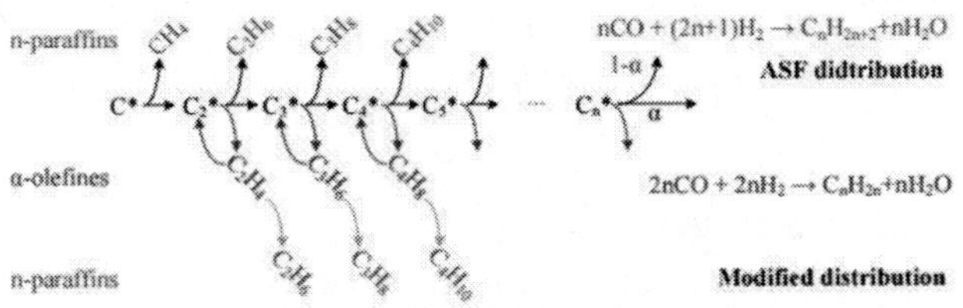

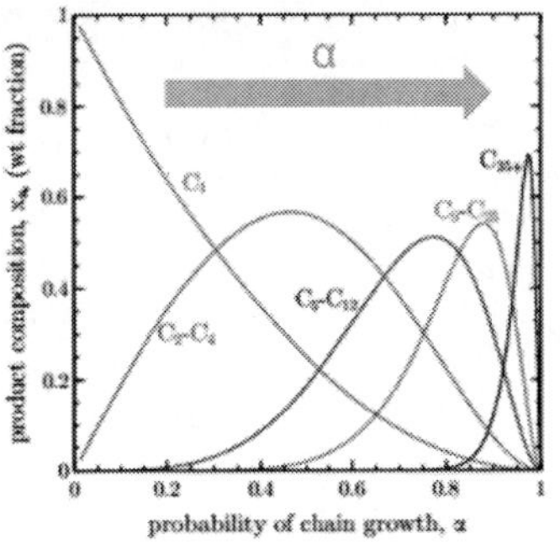

- **Low Temperature FTS (200-240 °C)**
  - Co
  - Wax, heavy hydrocarbons
  - Low water gas shift activity

Source: F. Kapteijn, R.M. De Deugd, J.A. Moulijn, Fischer-Tropsch synthesis using monolithic catalysts, Catalysis Today. 105 (2005) 350–356.

πδσωλ

GTL Plants

Existing GTL plants[1]

Location	Operator	Start-up	Capacity (b/d)	Note
Sasolburg, South Africa	Sasol	1955	5,600	Switched to nature gas in 2004
Mossel Bay, South Africa	PetroSA	1992	36,000	
Bintulu, Malaysia	Shell	1993	14,700	
Ras Laffan, Qatar	Oryx GTL	2006	34,000	Owned by Qatar Petroleum and Sasol
Ras Laffan, Qatar	Shell	2011	140,000	Owned by Qatar Petroleum and Shell
Escravos, Nigeria	Chevron	2013	34,000	Owned by NNPC and Chevron

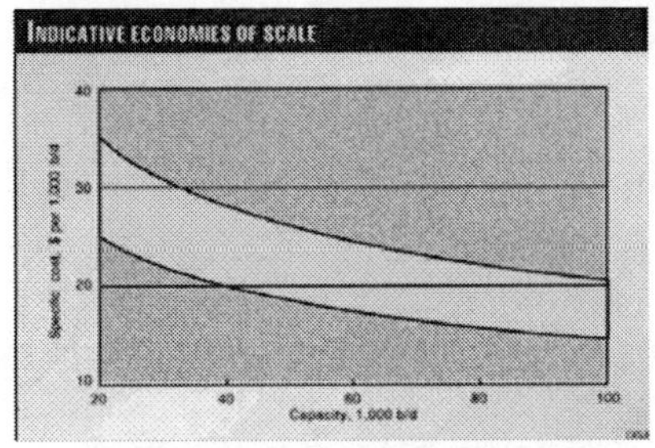

Product cost vs plant capacity.[2]

Less than 10% of the world's gas fields are large enough to sustain a 10,000+ bpd Fischer-Tropsch GTL plant[3]

1. https://www.eia.gov/todayinenergy/detail.php?id=28752
2. http://www.ogj.com/articles/print/volume-96/issue-38/in-this-issue/petrochemicals/gtl-technologies-focus-on-lowering-costs.html
3. Abstract, Process Economics Program Report 247 B, Small scale gas-to-liquids technology

πδσωλ

Modular GTL Plants

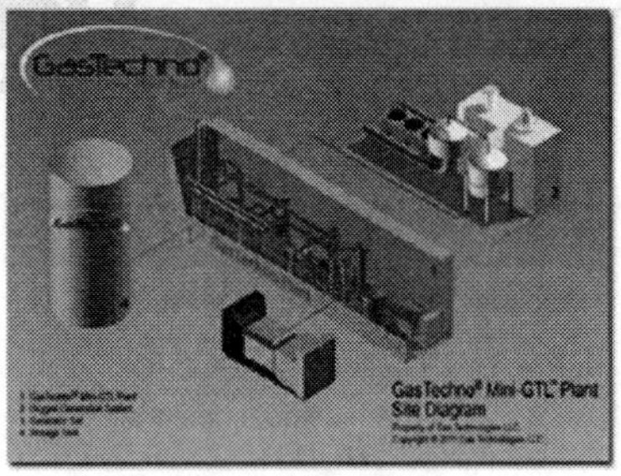

Modular GTL plant[1]

Numerous Companies Developing Mini-GTL Technology[2]

- **Reactor volume decrease**
- **High selectivity, Process equipment reduction**
- **Low capital investment, more profitable**

1. http://www.gastechno.com/gastechno-plant-scales.html
2. http://www.compactgtl.com/technology/petrobas

πδσωλ

Objectives

- **Reactor volume decrease**
  - One pot catalysts with high catalyst activity
- **Process equipment reduction**
  - High gasoline selectivity
- **This catalyst would require:**
  - High surface area
  - Excellent heat transfer
  - Improved metal dispersion
  - Presence of zeolite film to enhance secondary cracking and shape-selectivity
  - Mass transfer limitation release

Component	Function
Monolith	Rigid structure, high thermal conductivity
Cobalt-Al_2O_3	Syngas to hydrocarbons
Zeolite ZSM-5	Secondary acid cracking, shape selectivity

πδσωλ

Properties of Mesoporous ZSM-5

Meso-ZSM-5 created using the **desilication** technique

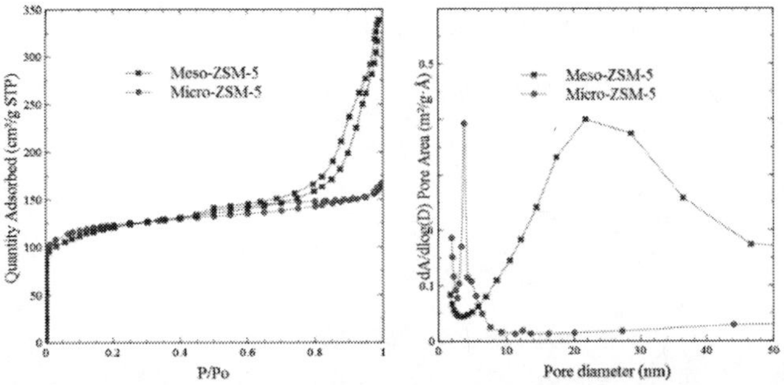

Sample	S_{BET} (m²/g)[a]	V_{micro}[b]	V_{meso}[c]	$D_{average}$[d]	ΔV_{micro}	ΔV_{meso}
Micro-ZSM-5	418.00	0.14	0.08	6.00	--	--
Meso-ZSM-5	421.00	0.10	0.36	15.00	-27.71%	353.09%

[a] Surface area obtained from Brunauer-Emmett-Teller (BET) measurements (S_{BET}). [b] t-plot micro-volume (V_{micro}). [c] BJH adsorption pore volume 1.7-500 nm (V_{meso}). [d] BJH adsorption average pore diameter($D_{average}$).

πδσωλ

Proposed FTS Catalysis

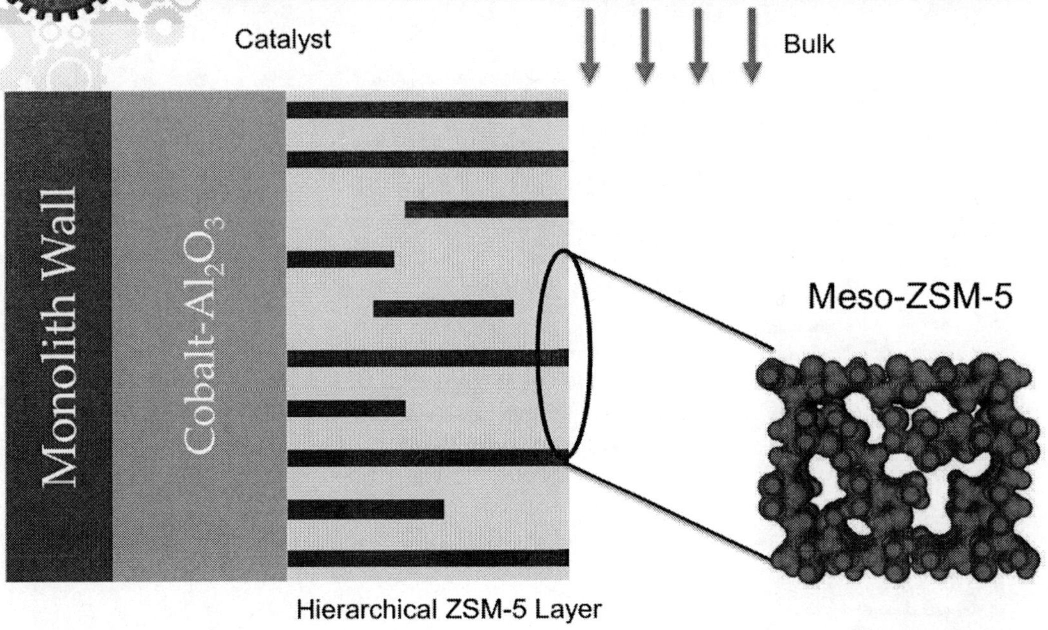

Proposed FTS Catalysis

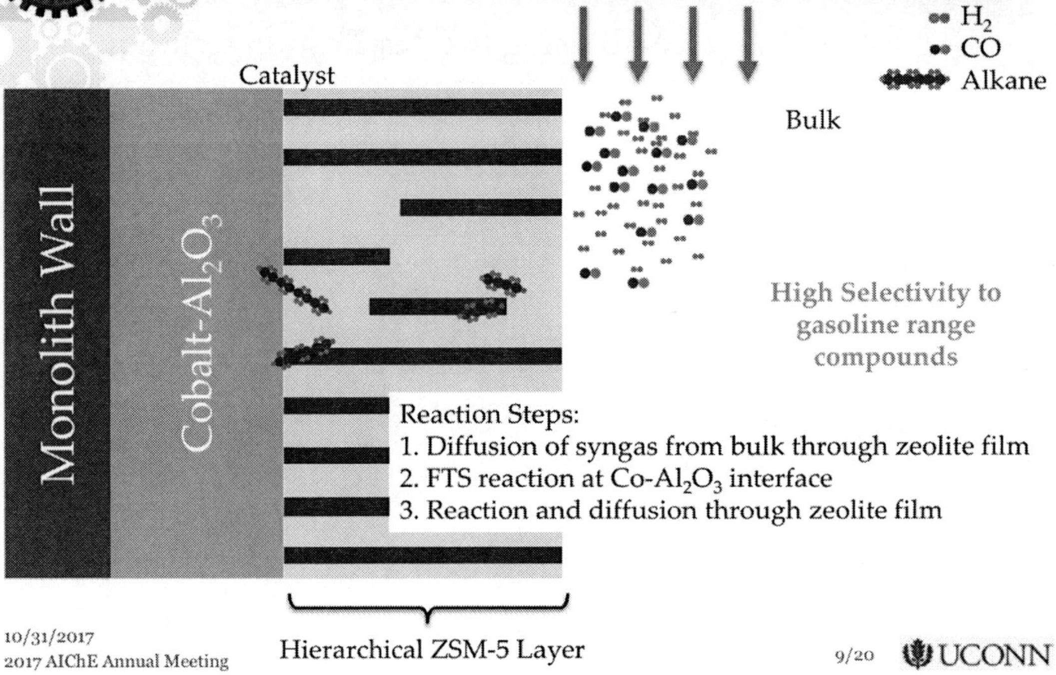

Reaction Steps:
1. Diffusion of syngas from bulk through zeolite film
2. FTS reaction at Co-Al_2O_3 interface
3. Reaction and diffusion through zeolite film

Catalysts Preparation Method

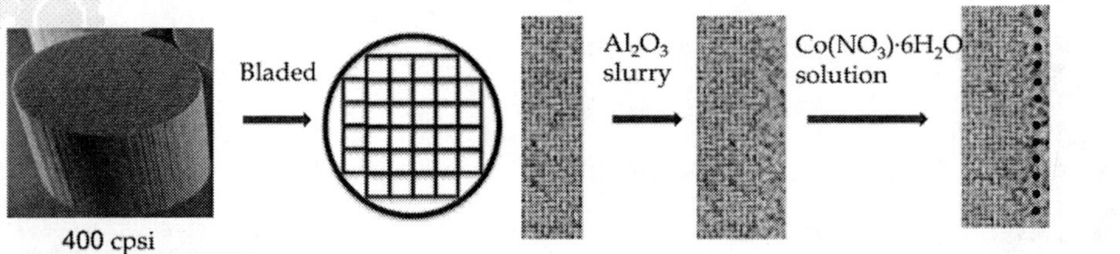

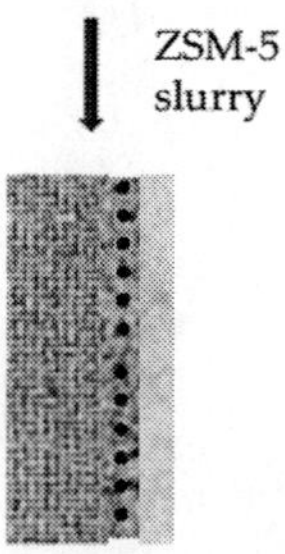

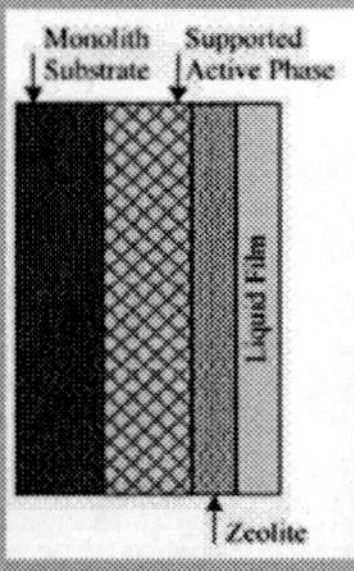

Figure 1. Schematic drawing of ZSM-5 coated monolithic catalyst

Catalysts were dried at 120 °C for 4 h and calcined at 400 °C for 12 h after each coating.

πδσωλ

Equipment Set-up

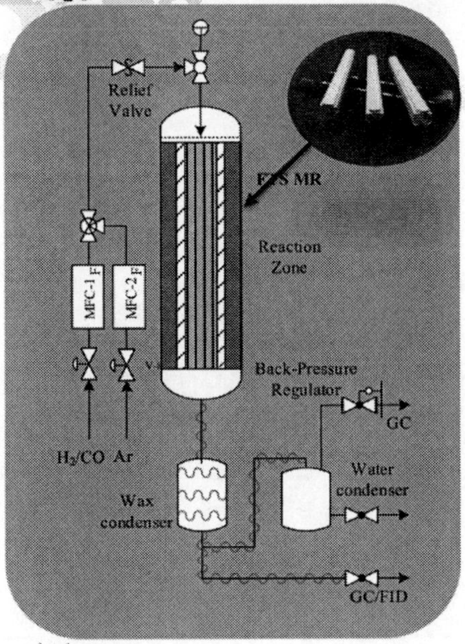

Catalysts Details

Catalyst	Catalyst weight(g)	Monolith (g)	Al_2O_3 (g)	Co_3O_4 (g)	ZSM-5 (g)
1.1 g-Micro	2.93 ± 0.03	1.42 ± 0.03	0.40 ± 0.08	0.13 ± 0.02	1.06 ± 0.04
1.1 g-Meso	3.11 ± 0.04	1.46 ± 0.03	0.50 ± 0.04	0.15 ± 0.02	1.07 ± 0.02
1.6 g-Meso	3.78 ± 0.01	1.45 ± 0.02	0.50 ± 0.01	0.16 ± 0.02	1.62 ± 0.03
1.9 g-Meso	4.01	1.43	0.53	0.16	1.89

- Premixed syngas (H_2:CO = 2:1).
- Ar was used as internal standard.
- Two-stage liquid trap. Wax trap was kept at 120 °C. Light oil and H_2O trap was at room temperature.
- Micro-GC 4900 with two columns and TCD for gases analysis.
- Agilent 6890 GC-FID with HP-5 column for wax and light oil analysis.

πδσωλ

FTS with Meso-ZSM-5 Coated Catalyst

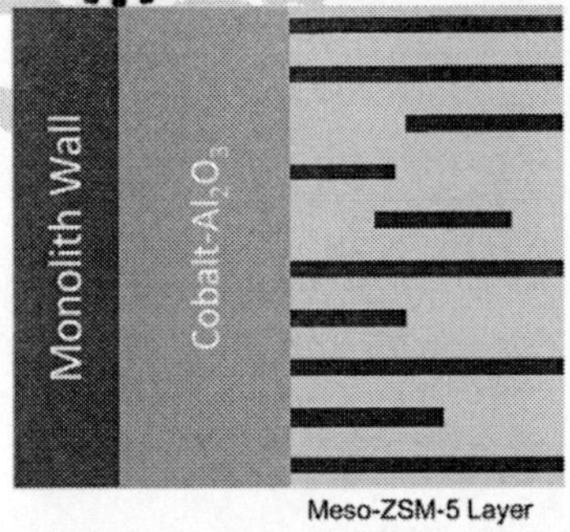

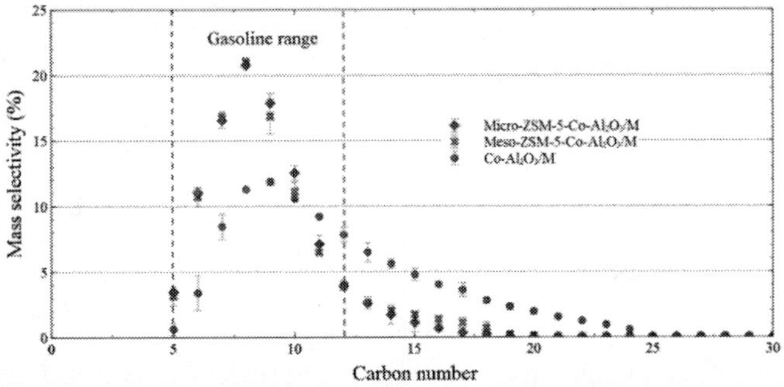

Component	CO Conv wt.%	Gas Selectivity wt.%	Liquid Selectivity wt.%	Mass Balance wt.%
$Co\text{-}Al_2O_3/M$	82	31.8	68.2	96
$ZSM\text{-}5/Co\text{-}Al_2O_3/M$	79	20.2	79.8	96
$Meso\text{-}ZSM\text{-}5/Co\text{-}Al_2O_3/M$	88	27.6	72.4	95

FTS with Meso-ZSM-5 Coated Catalyst

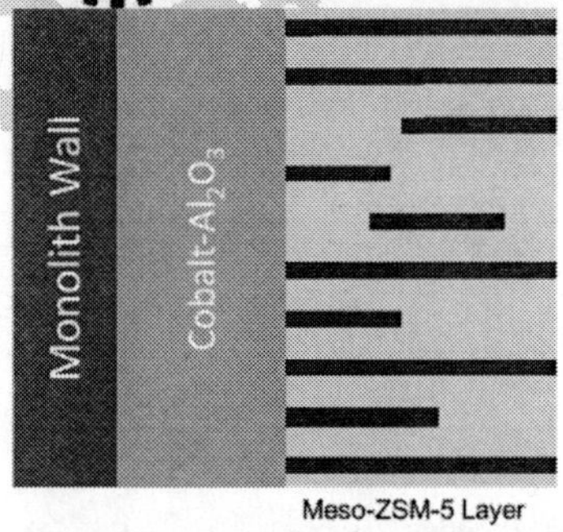

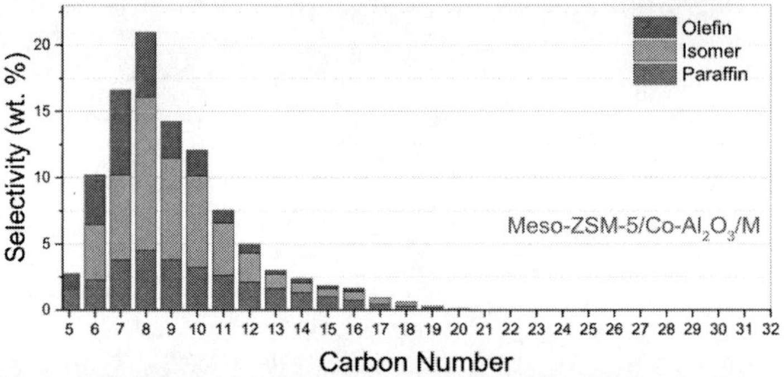

Component	Paraffin Selectivity wt.%	Isomer Selectivity wt.%	Olefin Selectivity wt.%
Co-Al_2O_3/M	69	26	5
ZSM-5/Co-Al_2O_3/M	29	50	21
Meso-ZSM-5/Co-Al_2O_3/M	29	47	23

Stability Comparison

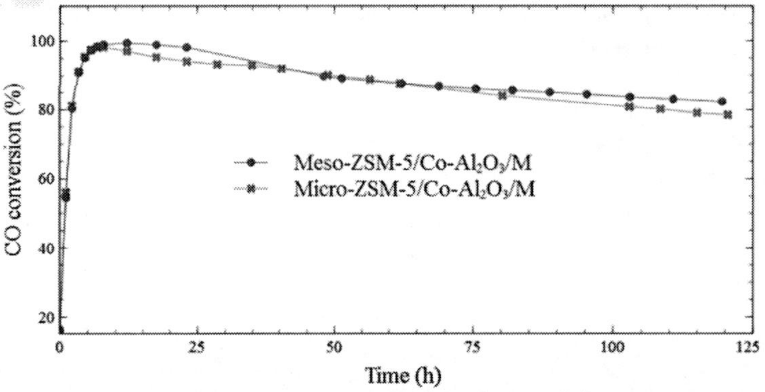

1.1 g-meso and 1.1 g-micro coated catalysts stability comparison within 120 h at 230 °C and 12 bar.

Catalysts coated with Meso-ZSM5 is more stable.

πδσωλ

Performance with Different Coating Thickness

Co-catalysts performance of 1.6 g meso-ZSM-5 at different pressures. Reaction condition: T: 230 °C; Syngas flowrate: 35 ml/min; Time on stream: 48 h

Catalyst	CO conv (%)	Carbon species selectivity (C%)			Oil phase selectivity (wt.%)		
		C_{5+}	C_5-C_{12}	C_{13+}	Paraffins	Isomers	Olefins
1.1 g-Meso	88.4 ± 0.6	77.3 ± 2.0	87.6 ± 1.3	12.4 ± 1.3	29.2 ± 1.8	44.9 ± 1.7	25.9 ± 0.5
1.6 g-Meso	87.8	80.2	93.2	6.8	27.5	50.6	21.9
1.9 g-Meso	85.1	81	92.9	7.1	26.1	50.2	23.7

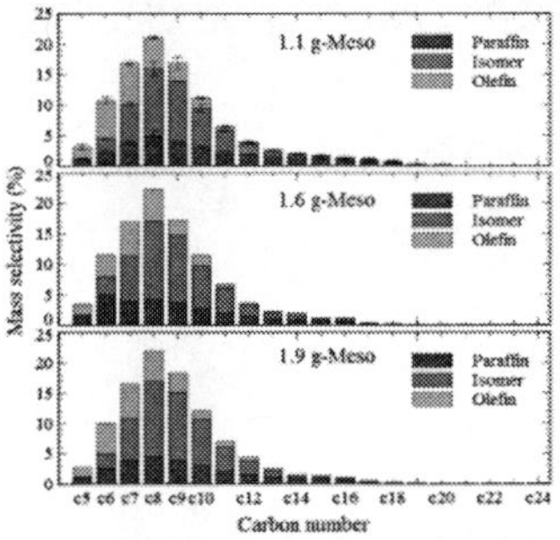

Liquid products distribution for different amount of meso-ZSM-5 coated co-catalysts.

- CO conversion didn't change too much.
- Gasoline selectivity was above 90 wt%.
- Isomers selectivity was elevated with ZSM-5 coating

πδσωλ

Performance at Different Pressures

Co-catalysts performance of 1.6 g meso-ZSM-5 at different pressures. Reaction condition: T: 230 °C; Syngas flowrate: 35 ml/min; Time on stream: 48 h

Catalyst	CO conv (%)	Carbon species selectivity (C%)			Oil phase selectivity (wt.%)		
		C_{5+}	C_5-C_{12}	C_{13+}	Paraffins	Isomers	Olefins
1.6 g -6 bar	55.3	84.6	88.9	11.1	33.1	44.3	22.6
1.6 g-12 bar	87.8	80.2	93.2	6.8	27.5	50.6	21.9
1.6 g-20 bar	92.8	78.4	92.4	7.6	28.1	50.1	21.8

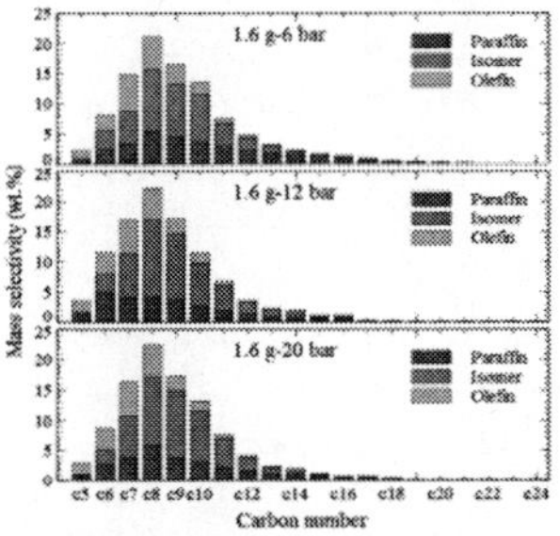

Liquid products distribution for 1.6 g meso-ZSM-5 coated co-catalysts at different pressures.

- CO conversion increased with increase of pressure.
- Gasoline selectivity reached highest at 12 bar.
- Isomers selectivity was elevated with ZSM-5 coating.

πδσωλ

Conclusions

- **Supported Zeolite/Co-Al_2O_3/monolith catalysts were created, characterized and evaluated for FTS**
- **Catalyst coated with Meso-zeolite ZSM-5 exhibits:**
  - High CO conversion
  - Excellent selectivity to gasoline range (C_5-C_{12}) compounds
  - Improved conversion to higher RON olefins and isomers
- **Recommended conditions for catalysts:**
  - Moderate loading of Meso-ZSM5: 1.6 g
  - Moderate pressure: 12 bar

πδσωλ

Acknowledgements

ACS-PRF Award No. 53648-DNI5

UConn Center for Clean Energy Engineering

Thank you!

Department of Chemical & Biomolecular Engineering
University of Connecticut

http://pdsol.engr.uconn.edu

10/31/2017
2017 AIChE Annual Meeting

19/20 UCONN

499690 Valorization of Natural Gas Liquids from Shale Gas

Abstract:

Recent shale gas boom has transformed the US energy landscape. Most of the major shale basins are located in remote locations and historically non-gas producing region. Therefore, many major shale basins regions are lacking the infrastructure to distribute the extracted gas into the rest of the US and particularly the Gulf Coast region. Here, we propose a new shale gas valorization process that converts natural gas liquids (NGLs) in shale gas into higher molecular weight hydrocarbon, which is more valuable to transport from these remote basins than NGL or its constituents such as ethane and propane.

In this process, natural gas liquid is transformed to higher molecular weight hydrocarbon. Compared with the existing technologies, our new approach has the following advantages: (1) This process minimizes the production of by-products. (2) This process is amenable to size reduction.

We demonstrate that, through a case study for a 96 MMSCFD shale gas feed from the Bakken region, the total capital cost for the proposed process is much cheaper than existing Gas-to-Liquid (GTL) processes such as Fischer-Tropsch and Methanol-to-Gasoline process. Also, we estimate the break-even price for our process to be lower than GTL-Fischer-Tropsch process. Therefore, this new process is an economically attractive step to upgrade shale gas in remote shale basins.

496044 Energy Sustainability Analysis of H_2 Production

Abstract:

The perspective for H_2 as an energy carrier to decarbonize energy services brings about a more and more relevant role for technologies of hydrogen production. As matter of fact, the yearly majority (>95%) of H_2 is produced from fossil fuels by steam reforming or partial oxidation of methane with only a small quantity by other routes, such as biomass gasification or electrolysis of water, hence the needs to explore new routes to produce H_2 appears evident. Stringent is the necessity to explore different technologies even at the infancy state, to produce hydrogen using different feedstocks. In this respect the energy analysis of H_2-producing technologies aimed to evaluate the net energy produced or Energy Return Of Investment (EROI) and Energy Payback Time (EPT) in order to score them towards energy sustainability is utmost of importance. To this end, the present paper is aimed to test three of the most advanced processes: *i)* steam reformer of methane, *ii)* electrolysis of water by using photovoltaic (PV) electrical energy and *iii)* dark fermentation (DF) of organic refuses (OR) by mixed microorganisms Hydrogen Producing Bacteria (HPB) mainly *Clostridium spp.* The three technologies were tested at laboratory scale for about one year, including the steam reformer owing the necessity to test the hypothesis to produce hydrogen at "the point of use" rather than trying to distribute it. In the panorama of bioH_2 production, which include heterotrophic and autotrophic microorganisms, the choice of DF is based on many reasons: *i)* reactions do not require light, so hydrogen is produced throughout the day and night, *ii)* the liquid end products of DF can be used as feedstock for Anaerobic Digestion (AD) to produce CH_4 with additional energy produced, and *iii)* because the OR are so abundant (> 70 % of wastes) and almost uniformly distributed.The data acquired by experimental campaigns were used to evaluate and scoring the energy sustainability of the three process by the application of a previous candidate procedure (Di Addario et al.,2016). The procedure is recognised into three different steps: a first screening is performed by using the Energy Sustainability Index (ESI), which takes into account the total Produced Energy under form of hydrogen referred to the direct spent energy (heat and electricity). The second step is the Analogical Model, i.e. the quantification of each chemicals, materials and energy to estimate the Useful Energy with a Life Cycle Approach (LCA), taking into account all the Indirect Energy terms. At such a step the energy embedded in the feeds, i.e. methane and water were evaluated for the steam reformer and the electrolysis process, respectively. For the DF process instead, the energy embedded in OR was assumed to be equalto zero, following the allocation procedures in LCA analysis. The third step is the evaluation of two parameters: EROI and EPT, considering all the involved energy flows. The procedure is anuseful tool to score the three processes towards sustainability: higher score correspond to high EROI & low EPT, and this permits to select the most sustainable choice among the three analyzed H_2-producing processes from an energy point of view.

Di Addario, M., Malavè, A.C.L., Sanfilippo, S., Fino, D., Ruggeri, B., 2016. Evaluation of sustainable useful index (SUI) by fuzzy approach for energy producing processes. *Chemical Engineering Research and Design*, 107, pp.153–166.

501652 Chemical Looping Based Technology for High Efficiency Production of H_2 from Ammonia (NH_3)

Abstract:

Hydrogen (H2) is widely considered to be an integral part in solving the issues surrounding the consumption of fossil fuels. As a carbon neutral fuel, H2 can be used as an energy source with minimal emissions for a wide range of uses including chemical production and transportation. One of the most advantageous properties of using H2 as a fuel source lies in its high specific energy of 39.8 kWh/kg (HHV). Despite the advantages offered by the widespread utilization of H2, market penetration is prohibited by the high transportation costs associated with its low volumetric energy density of 1.55kWh/L and thus high compression losses in the storage of H2. In an effort to overcome these limitations, the use of ammonia (NH3) as a hydrogen carrier to minimize transportation and storage costs has been proposed. This vision for improving the market penetration of H2, is contingent upon development of high efficiency H2 from NH3 processes. OSU has developed a chemical looping based NH3 to H2 (ATH) process that can operate at high thermal and H2 production efficiencies. The OSU chemical looping approach consists of two reactors to convert NH3 to H2. The first reactor, namely the reducer reactor utilizes an intrinsic O2 gradient driven by the reduction potential of NH3 and a metal oxide like Fe3O4, ZrO2, WO3 to efficiently crack NH3 to a mixture of N2, H2 and H2O. The reduced metal-oxide from the reducer reactor is re-oxidized in an oxidizer using H2O as the oxygen source. This ensures ~1.5 moles of H2 production per mole of NH3 (nearly 100% H2 production efficiency). This presentation will initially focus on the thermodynamic rationale for developing a chemical looping based NH3 to H2 process. This will be followed by the heat balance approach to achieving the maximum H2 production efficiency while maintaining a high energy efficiency values. The relevant experimental data (using Thermo-gravimetric Analyzer and a fixed bed reactor) that identifies key design conditions associated with the kinetics of the reducer and the oxidizer reactor will be presented. A series of directed sensitivity studies that identify the trends associated with the chemical looping system and the design parameters that need to be de-risked in further scale up of this technology will be discussed.

496846 Distributed Ammonia Manufacturing

Abstract:

To eliminate the dependency on fossil fuel and the carbon dioxide emissions, carbon-free ammonia can be synthesized in small wind-powered plants, and from totally renewable resources: hydrogen from electrolysis of water and nitrogen from pressure swing adsorption of air. We are making carbon-neutral ammonia in the UMN-WCROC plant, located in the agricultural region of Morris MN.[1] While this process is working, the ammonia costs twice that made conventionally, which is due to downscaling (40,000 scaled down process.) To overcome the difficulties associated with the energy demand for ammonia synthesis, and to improve the economics of the small plant, we are seeking a more efficient process to synthesize carbon-neutral ammonia.

One strategy to improve this process is to design a new class of materials for the catalysis, that allows synthesis at lower temperature and pressure. However, a century of research confirms that this is not trivial and immediately viable.[2] Essentially with the conventional catalyst, high temperature is a must in order to obtain fast kinetics, while high pressure ensures the conversion. If we can efficiently remove low concentrations of ammonia formed in the reactor, then we can keep the reaction far from equilibrium, and intrinsically we obtain rates closer to forward reaction rates. Thus the constraints of equilibrium is surpassed. We have proposed the application of solid absorber for high temperature ammonia separations.[3,4] In this process, ammonia can be synthesized at much lower pressure, with rates comparable to high-pressure conventional processes. The capital and operating costs of low pressure reaction-absorption process is improved since the work of compressing the fresh feed is significantly reduced, and the need for heat exchange also drops.

The low pressure reaction-absorption process suggests a strategy for making ammonia fertilizer locally, using wind-generated electricity. Alternatively, this ammonia made in rural area, can be used as a fuel, or a hydrogen carrier for fuel cell use in more urban areas.

References:

(1) Reese, M.; Marquart, C.; Malmali, M.; Wagner, K.; Buchanan, E.; McCormick, A.; Cussler, E. L. Performance of a Small-Scale Haber Process. *Ind. Eng. Chem. Res.* **2016**, *55* (13), 3742.

(2) Schlögl, R. Catalytic Synthesis of Ammonia—A "Never-Ending Story"? *Angew. Chemie Int. Ed.* **2003**, *42* (18), 2004.

(3) Malmali, M.; Wei, Y.; McCormick, A.; Cussler, E. L. Ammonia Synthesis at Reduced Pressure via Reactive Separation. *Ind. Eng. Chem. Res.* **2016**, *55* (33), 8922.

(4) Wagner, K.; Malmali, M.; Smith, C.; McCormick, A.; Cussler, E. L.; Zhu, M.; Seaton, N. C. A. Column Absorption for Reproducible Cyclic Separation in Small Scale Ammonia Synthesis. *AIChE J.* **2017**.

Towards Bio refinery Process Integration in Oil Refineries

Harvey Arellano-Garcia*, Elham Ketabchi, Tomas Ramirez Reina

Department of Chemical and Process Engineering, University of Surrey, Guildford, GU27XH, United Kingdom

**h.arellano-garcia@surrey.ac.uk*

Abstract

An investigation into the integration of bio-refineries with oil refineries is considered in this study. The identified possible routes of integration will be discussed in this contribution. For this purpose, the different proposed alternatives and their configurations were simulated and analysed. These flow schemes formed the basis upon which a series of simulations in CHEMCAD™ and Aspen plus® were undertaken. The developed models simulated key integrations e.g. a gasification unit that is fed from pyrolysis oil, biodiesel, and refinery residue, before being combined into one system involving all three. Varying forms of synthesis for these three feeds were also considered, focusing on novel techniques as well as environmentally friendly options that made use of waste products from other processes. The simulations found that the resulting flow from the gasification unit contained the expected composition of H_2 and CO_2, with CO in slight excess. The H_2:CO ratio was also found to be adequate enough for further processing in the form of producing hydrogen (Water Gas Shift reaction) and liquid hydrocarbons (Fischer-Tropsch). This study also found that through the manipulation of oxygen content, the H_2:CO ratio was improved. Changes in composition of the output from the gasification unit were also evaluated through influencing the input flows from the three sources mentioned earlier. Furthermore, through the use of low value streams and recycled waste, as well as the use of non-fossil based fuels, this integration drastically reduces the negative environmental impact that the refinery has as well as preventing excessive wastes of energy, while not increasing costs for advanced equipment or pre-processing.

Keywords: Process integration, bio-refinery, oil refinery, GHG, waste reduction.

1. Introduction

The main idea is to create a flow sheet of integrated units and processes of an oil refinery with a bio-refinery. The flow sheet construction has been inspired by the work of Schinski (2014). This integration is appealing as bio-refinery feed stocks are more readily available due to the wide variety of biomass that can be used, i.e. sugar and wood that produce bio-fuels, pyrolysis oil in this study, of comparable quality to fossil fuels and therefore, energy. Furthermore, bio-diesel produced from waste water treatment as an extra added-value product has been considered (Kiracofe, 2010).

These processes mentioned including oil refinery residue, oxygen and steam are all inputs for the gasification unit in order to produce syngas.

Once the syngas has been purified, it enters the WGS and FT reactors to produce fuel (liquid hydrocarbon) and hydrogen. Energy obtained from these processes can also be used in a combined heat and power (CHP) unit to maximise the process integration.

The tail gas outlet from the FT reactor enters a separation unit, separating the carbon monoxide from the stream. The remainder of which, enters the refinery again.

Pyrolysis oil used in this study is specifically fractionated butanol that comes from sugar fermentation. This is more favourable as a bio-fuel due to the need of mitigating greenhouse gas (GHG) emissions with the use of developing technologies, when compared to conventional fuel. Figure 1 shows the flow sheet used in this work for modelling and simulations in CHEMCAD™ and Aspen Plus®. Moreover, sensitivity analysis was carried out to determine sensible output changes by varying several input properties.

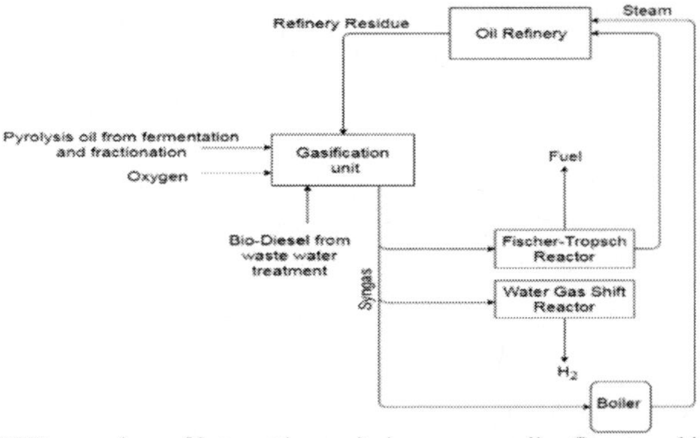

Figure 1. BFD overview of integrating units/processes: oil refinery and bio-refinery

The other route considered in this flow sheet is the production of bio-diesel, as seen in the bottom of Figure 1, from the waste water sludge that is produced from the waste water treatment process via trans-esterification, then using this bio-diesel as another input for the gasification unit. The waste water sludge is rich in fatty acids, which are the main components required for trans-esterification that produces bio-diesel and glycerol. These two products are easily separated through methods such as gravity settling (Kiracofe, 2010).

The gasification unit operating under isothermal and steady state conditions also uses refinery residue, in addition to the bio-diesel and pyrolysis oil, with oxygen and steam to produce syngas, which is the main product, while also producing slag. The use of these three main streams that originate from previously unused material, to create syngas, is to promote more "green" energy production and to offer solutions to dwindling fossil fuel supply.

The conversion of biomass via gasification and hydro-pyrolysis followed by the conversion of natural gas via FT synthesis to liquid fuel is another alternative to these integration pathways proposed by Gencer et al (2014) and is included it in the integrated approach. The tail gas produced from the FT reactor could also be entered in a separator, where carbon dioxide from the remainder gas is separated before being returned to the oil refinery. Another aspect of integration proposed is the conversion of syngas to steam via boiling to produce steam for the CHP plant. The output of the boiler being steam could either be directly used in the refinery operations or could be converted to electricity for the benefit of a refinery.

2. Design and simulation

2.1. Mass and energy balance of the gasification unit

The first step is to identify the specifications of the inputs to the gasification unit being pyrolysis oil, biodiesel, and refinery residue. It is considered that the product of a conventional pyrolysis unit with the efficiency of 20 % is 0.956 kg/s using information provided in the work by Desai (2011). The common composition of pyrolysis oil is obtained from Vivarelli and Tondi (2004). The bio-diesel used as an input in this flow sheet is obtained from the process of using waste from waste water treatment as mentioned before. The values used for this simulation are derived from Kiracofe's 2010 work as well as simple mass and energy balances leading to the values demonstrated in the caption for Fig 2. The composition of bio-diesel chosen for this work is derived from the work of Hoekman et al.(2012). The amount of bio-diesel shown here is minimal due to conversion processes only having yields of maximum 13 % as shown in Kiracofe's (2010) research. Considering the input volumes to an oil refinery, we can assume that the outlet volume in the form of refinery residue will be of the order of $1x10^4$ kg/h. In this case, the refinery residue was initially chosen to be 60,000 kg/h for use in the simulation. Standard stoichiometric approaches are not applicable for biomass processes due to the fact that they are not composed of well-defined chemical species, therefore, the amount of oxygen required for the gasification process was derived from the calculation of the empirical formula in (Bernocco et al., 2011). Moreover, through the use of simple energy balance, the temperature of the oxygen and steam mixture required for the gasification unit was calculated.

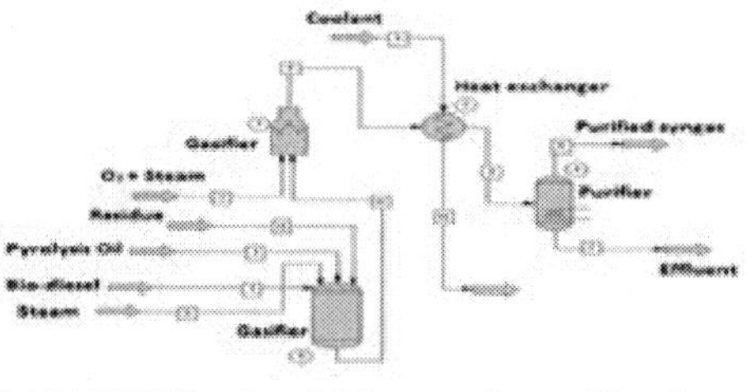

Figure 2. CHEMCAD™ Version 6.5.5 setup for gasification and purification (gasification conditions: 40 bar, 1400 °C, steam temperature 350 °C and 99.6 % O_2 concentration – adapted from Chao 2015)

The simulation was improved, however, by reducing the amount of oxygen for the gasification process in order to obtain higher quality syngas, with a more desirable ratio of $CO:CO_2$, in favour of CO.

2.2. Mass and energy balance of the water gas shift reactor

Following the gasification modelling (Fig. 1), the next step is the water gas shift reactor. The produced gas or the purified syngas enters the WGS reactor in order to produce hydrogen rich syngas. For this reaction, an equilibrium reactor is selected for the WGS reaction due to the chemistry involved dictating the reliance on equilibrium to favourably produce hydrogen.

Figure 3. Simulation setup for WGS reaction in Aspen Plus® Version 8.6 (WGS reactor conditions: 36.8 bar, outlet temperature 612 °C, Steam temperature 350 °C, Steam pressure 165 bar – adapted from Kopyscinski et al, 2010)

3. Results and Discussion

The composition products of the WGS reactor are presented in tables 1 and 2. The values demonstrate high amounts of carbon monoxide followed by hydrogen and carbon dioxide with the percentages of approximately 4 %, 95 %, and 1 %, respectively. The obtained composition is as expected, showing the values for good quality syngas which has been improved even more by conducting sensitivity analyses. It must be noted that the amount of methane produced in the gasification unit were negligible. As can be seen in table 2, the WGS reaction products obtained, result in having the desired ratio of 2.0105 being suitable for further processing such as FT reactions.

Table 1.Syngas composition obtained from simulation after improvements made through sensitivity analyses

Syngas composition	Amount (kg/h)	Amount (kmol/h)
H_2	3564.34	1768.20
CO	61817.50	2206.98
CO_2	60.33	1.37

Table 2.WGS reaction (WGSR) products derived from simulation

WGSR Product Composition	Amount (kmol/h)
CO	1231.06
H_2	2475.05
H_2O	2376.84
CO_2	827.84

4. Conclusions

The results of the simulation imply that the syngas produced from the gasification process demonstrate the expected composition of CO, H_2, and CO_2 having high amounts of CO.

The results also depict the H_2:CO ratio of the WGSR products to be as expected, and thus, meeting the requirement for further processing. This ratio has been improved through enhancements made in the amount of oxygen, supported by using information obtained through sensitivity analysis. Moreover, changes in composition of the output from the gasification unit were also evaluated by modulating the input flows.

In conclusion, through the use of non-fossil based fuels as well as low-value streams and by recycling waste, this integration method will be able to decrease the negative environmental impacts of conventional oil refining while preventing excessive wastes of energy without increasing costs for advanced equipment or pre-processing.

Finally, it could be shown that these integrations will increase the profitability of any refinery that incorporates them properly and effectively while reducing overall refinery wastage.

References

Bernocco, D., Greppi, P., & Arato, E. (2011). *A novel approach to the biomass pyrolysis step and product lumping*. *Computer Aided Chemical Engineering* (Vol. 29). Elsevier B.V. http://doi.org/10.1016/B978-0-444-53711-9.50023-7

Chao, L. I., Zheng-hua, D. A. I., Ji, Y., Guang-suo, Y. U., & Fu-chen, W. (2015). Modelling and energy analysis of an integrated coal gasification and pyrolysis system for synthetic natural gas. *Journal of Fuel Chemistry and Technology*, *43*(7), 779–789.

Desai, P. D. (2011). Design of pyrolysis reactor. Retrieved June 28, 2016, from https://www.scribd.com/doc/100920538/Design-of-Pyrolysis-Reactor

Gencer, E., Mallapragada, D., Tawarmalani, M., & Lafayette, W. (2014). Synergistic Biomass and Natural Gas Conversion to Liquid Fuel with Reduced CO 2 Emissions. In *e 8th International Conference on FOCAPD* (pp.525–530).

Hoekman, S. K., Broch, A., Robbins, C., Ceniceros, E., & Natarajan, M. (2012). Review of biodiesel composition , properties , and specifications. *Renewable and Sustainable Energy Reviews*, *16*(1), 143–169.

Kiracofe, N. L. (2010). *Converting wastewater treatment facilities into biorefineries: Biodiesel from wastewater microorganisms*. University of Delaware.

Kopyscinski, J., Schildhauer, T. J., & Biollaz, S. M. A. (2010). Production of synthetic natural gas (SNG) from coal and dry biomass – A technology review from 1950 to 2009. *Fuel*, *89*(8), 1763–1783.

Schinski, W. L. (2014). Hybrid refinery for co-processing biomass with conventional refinery streams. United states: United states Patent and trademark office.

Vivarelli, S., & Tondi, G. (2004). *Bioenergy for a sustainable development*. Chile.

501634 Process Intensification in the Synthesis of Tributyl Citrate: Pilot Scale Validation

Abstract:

Over the few last decades, bio based plasticizers have been their market share due to the generalized toxicity concern from phthalate esters and other traditional plasticizers. Tributyl citrate (TBC) may be synthesized from citric acid and n-butanol in presence of an acid catalyst. Current production process is highly energy intensive due to its semi-batch configuration and the excess of butanol used at the reaction stage which must be recovered downstream.

The present work evaluated the production of TBC via reactive distillation and other intensifies configurations to increase energy and mass efficiency and furthermore profitability. The thermodynamic model to describe the non-ideal behavior of the reactive system has been regressed from liquid-liquid, solid-liquid , and vapor-liquid experimental equilibria of binary and ternary mixtures. The kinetic model has been regressed from experiments with homogeneous and heterogeneous catalysts. The conceptual design has been carried out using reactive residue curve maps with the introduction of the substitution degree to visualize multicomponent mixtures on a 2D color-scale. Pilot experimental evaluation was carried out in a 7m tall 3â€ wide stainless steel reactive distillation column available in our laboratory with structured packing similar to KATAPAK SP-11 with Amberlyst 70 as catalyst in the reactive section and 1â€ nutter rings in the stripping and recovery sections. The experimental pressure was 74kPa, the atmospheric pressure in Bogota, Colombia. Variables of study during the pilot experimentation include: reflux ratio, reactive mixture feed, use of pure butanol in a side stream, use of azeotropic grade over dehydrated butanol, and the use of prereactor. Pilot plant data were reconciled to develop a robust model for process up-scaling. An industrial process was modelled and optimized for a production capacity of 1000Ton/yr with a natured inspired algorithm. Results from optimization indicate a citric acid conversion>99.9%, selectivity towards TBC >90%, energy savings of 27% and cost reduction of 13% over the conventional process.

491296 Process Intensification in Multicomponent Distillation

Abstract:

Process intensification (PI) is an emerging concept in chemical engineering which describes the design innovations that lead to significant shrink in size and dramatic boost in efficiency in a process plant. Distillation, which accounts for 90-95% of all separations in a chemical plant and ~3% of the world's energy consumption, is therefore a crucial component in PI. In the context of multicomponent distillation, we believe that PI stands for minimizing the number of pieces of equipment as well as total cost while maintaining high efficiency through innovative process synthesis. This will lead to compact, easy-to-operate, energy efficient and cost effective processes. Here, we consider several different levels of PI in multicomponent distillation, including: 1) Introducing thermal couplings that eliminates some or all reboilers and condensers to save energy and reduce capital cost; 2) Using multi-effect distillation along with converting thermal couplings to liquid-only transfers to further reduce heat duty, enhance energy efficiency, and improve operability of thermally coupled distillation configurations; 3) Performing simultaneous heat and mass integration among thermally coupled distillation columns to reduce the number of columns and heat duty requirement; 4) Conducting any thermally coupled distillation in only one single column shell using the form of a dividing wall column that is fully operable; 5) Being able to systematically derive numerous configurations that use 1 to (−2) column shells from a traditional (−1)-column configuration. Here, we demonstrate these aspects of PI through examples, illustrating the importance of PI in multicomponent distillation and how this concept can be utilized by process engineers to design cost effective and energy efficient distillation systems.

489897 Analysis of Key Metrics in the Use of Oxidative Dehydrogenation for the Production of Ethylene from Shale Gas Ethane

Abstract:

The world-wide market for primary light olefins will approach half a trillion pounds per year by the year 2020. Light olefins are conventionally produced in the US by the steam cracking of hydrocarbons from ethane to gas oil. With the abundance of shale gas components, the feedstock of choice in the US is ethane. The reaction technology is decades old and results in a wide range of byproducts, some valuable, some with fuel value, and some requiring waste treatment before disposal. A techno-economic analysis was performed using reaction data provided by the DOE-INL for Oxidative Dehydrogenation (ODH) and modeling the entire expected processing using ASPEN-HYSYS V. 9.0. A set of performance metrics and a cost of production procedure commonly called a required netback analysis were used to compare this intensified ODH process with conventional steam pyrolysis technology. The new ODH route uses a mixed metal oxide catalyst in packed tubes with a heat transfer fluid in the reactor shell for heat management. Results indicate that the new route has much lower variable cost, fixed costs and capital recovery reaching shutdown economic level. Regarding the 20+ metrics developed, the ODH route shows significant advantages and integrates well with certain novel separations technology currently in the pipeline.

497448 Continuous Manufacturing As an Enabler for Intensified Process Design and an Integrated Control Strategy in Pharmaceutical Synthesis

Abstract:

Development, scale-up, and implementation of a continuous multi-step route for manufacturing of an active pharmaceutical ingredient are detailed. Long recognized as essential for economical process intensification in other chemical industries, continuous manufacturing stands to offer many of the same advantages to the pharmaceutical industry. These include more inherent scalability on account of improved heat and mass transfer, access to more aggressive reaction conditions, smaller footprint equipment sets placing less material at-risk, more opportunities to leverage online measurement and automation, and established control of the process at steady-state. The integration of all of these advantages enables implementation of a control strategy that ensures quality under more cost-effective, and often safer, flow processing conditions.

For the case study herein, the transition from laboratory development to a continuous process demonstration operating at 50â€“100% of the commercial-scale in walk-in fume hoods is described. Scale-up efforts integrating multiple unit operations in flow and leveraging process analytical technology (PAT) are demonstrated to test process robustness and generate process knowledge that directly informs the target conditions and operating ranges for the commercial-scale manufacturing process. In this example, transfer of the process for a late-stage clinical supply campaign within the external manufacturing network is detailed, culminating in a continuous route with two plug flow reactor (PFR) transformationsâ€”an amidation and a high-temperature deprotectionâ€”operating in conjunction with a continuous mixed-suspension, mixed-product-removal (MSMPR) crystallization and parallel filtration with *in situ* solids dissolution. Genotoxic impurity control is afforded through process design and engineering controls and is verified with online and offline analyses. The process is shown to operate in continuous mode for 18 days, affording more than 150 kg drug substance in exceptional purity. These steps will be at the core of a multi-step continuous unit operations sequence in the commercial process.

493712 A Systematic Method for Chemical and Biochemical Sustainable Process Synthesis, Design and Intensification

Abstract:

In this highly competitive world of chemicals, challenges like ever-increasing energy and utility cost, pressure to increase resource efficiency and new environmental regulations [1] requires systematic process synthesis-design solutions. The objective of systematic process synthesis-design is to identify the processing route, which gives a desired product from a specific set of raw materials including unit operations design, utility needs, waste generation and sustainability parameters. However, to find innovative and more sustainable process design solutions, extension of the current methods is necessary that can be generated by process intensification. Process Intensification (PI) provides a pathway to design new and innovative solutions or retrofit of existing ones leading to economic, sustainable and efficient solutions. Thus, it is logical to integrate process synthesis-design and intensification to generate more innovative, sustainable and efficient chemical and biochemical solutions.

In this work, a generic three-stage hierarchical decomposition method based on unit operation, task and phenomenon level is applied for sustainable production of succinic acid. This approach examines several succinic acid processing routes from different feeds including bio (glucose, glycerol etc.) and petro (maleic anhydride) based raw materials that are collected through available data and current technologies reported in literature. Using this information, a generic superstructure of processing routes is created that describes a network of configurations representing multiple designs for the production of succinic acid. The next step is to solve the superstructure in order to find the optimal processing route or flowsheet. This has been done through Super-O [2], a software interface that guides the formulation and optimization of the superstructure to find optimal processing route. Thus, in stage 1, process synthesis is performed to generate an optimal flowsheet among numerous alternatives for succinic acid. In stage 2, the base-case design is developed based on the route selected in stage 1, which is then further analyzed to identify the process hotspots. This is done by first performing simulation to obtain the steady state mass and energy balance data for the selected route. These data are then used for economic analysis, sustainability evaluation [3] and life cycle assessment [4] where a set of calculated indicators are used to identify the bottlenecks or process hotspots. These hotspots are then translated to design targets that are used in the next (3rd) stage where, a phenomenon based synthesis-intensification method is applied [5]. Here, the phenomena are identified by using the PI knowledge base that are representing the tasks performed by unit operations within the base case design. These phenomena are further analysed and re-combined using combination rules to generate new and/or existing unit operations including hybrid operations which in turn are combined to form new and innovative process alternatives. Finally, from the set of feasible intensified process alternatives for production of succinic acid, the best in terms of economic and environmental sustainability is identified.

In this presentation, the implementation of 3-stage method into an integrated multi-scale computer-aided framework will be presented and highlighted by its application to the production of succinic acid to generate more sustainable solutions. The presentation will also discuss the computer aided tools used in this work to achieve the sustainable production of succinic acid.

References:

[1] Chemicals, Retrieved April 4, 2017, from https://ec.europa.eu/growth/sectors/chemicals_en

[2] M. Bertran, R. Frauzem, A. Arcilla, L. Zhang, J. M. Woodley, R. Gani, Computers & Chemical Engineering, 2017, doi: 10.1016/j.compchemeng.2017.01.030.

[3] A. Carvalho, H. A. Matos, R. Gani, Computers & Chemical Engineering, 50 (2013): 8-27.

[4] S. Kalakul, P. Malakul, K. Siemanond, R. Gani, Journal of Cleaner Production, 71 (2014): 98-109.

[5] D. K. Babi, J. Holtbruegge, P. Lutze, A. Gorak, J. M. Woodley, R. Gani, Computers & Chemical Engineering, 81 (2015): 218-244.

489611 Strategies for Process and Size Selection of Natural Gas Liquefaction Processes: Specific Profit Portfolio Approach By Cost Based Optimization

Abstract:

The natural gas liquefaction process is highly energy intensive process due to the cryogenic operating characteristics.[1] Thus, the natural gas liquefaction process has the energy intensive characteristics because of its cryogenic operation condition and most of energy is consumed by compression unit.[2] Therefore, minimizing compression energy requirement is the major concern in design and optimization of the natural gas liquefaction process. The natural gas liquefaction processes can be classified by the number of refrigeration cycles and the type of refrigerants used.[3] Because different processes have different characteristics in energy efficiency and number of equipment, the comparison between various liquefaction processes have to be performed to find the optimal design; in terms of the energy as well as cost Furthermore, the plant size also affects to the capital cost. In economics of scale, larger size of plant can make more profit. The amount of the increased profit by increasing the size is different to different processes. Therefore, the most profitable process can be different by different size of plant. For these reasons, it is needed to optimize in terms of the cost with different liquefaction processes to find optimal design and operating condition in various plant sizes.

The main purpose of this study is to make strategies of process and size selection by finding the optimal natural gas liquefaction processes in various plant sizes. To find optimal design and size, the operating cost including energy requirement and capital cost have to be considered simultaneously. The cost based model can consider the plant size which cannot be considered in thermodynamic model. Moreover, the price and flow rate of the natural gas feed and LNG also affect to the profit. Therefore, the profit optimization model have to be developed because it can consider the cost of law material and the product.

The profit optimization model is developed with energy self-sufficient liquefaction system. Among various natural gas liquefaction processes, Single Mixed Refrigerant (SMR), Dual Mixed Refrigerant (DMR), and Propane-precooled Mixed Refrigerant (C3MR) processes are targeted because they are well known and have representative refrigeration characteristics. Optimization model is developed by gPROMS software and the model includes thermodynamic model and cost model. The cost model basically follows the six-tenth factor rule. The objective function is max. specific profit which is profit per unit mass of LNG (USD/ton-LNG). The unit prices of natural gas and LNG are applied based on the Henry-hub prices announced by EIA, the unit prices of the natural gas and the LNG are applied 2.5 USD per Mcf and 4.5 USD per Mcf respectively to reflect recent prices.[4] The detailed mathematical model basically follows the model which is presented in previous work.[5]

The profit optimization with various size of SMR, DMR, and C3MR processes are performed and the results of optimizations are compared and discussed. The portfolios for the liquefaction ratio, specific energy requirement, specific cost, and specific profit obtained by the profit optimizations for each process. The most economical size range for each process design is figured out by drawing specific profit portfolios. SMR process which has the simplest configuration has an economic strength in the small plant size. Even though the energy efficiency is the lowest among three liquefaction processes, the low plant cost of SMR process can cover the disadvantage of the low efficiency in the small plant size. However, as the plant size increases, DMR process catches up SMR process in terms of the specific profit. The configuration of DMR process is more complex and it requires more equipment than SMR process. For the medium size of plant, the higher energy efficiency makes the DMR process economical compare to SMR process. For the large plant size, the C3MR process is most economical because of the highest efficiency although it requires much more equipment compare to SMR and DMR processes.

This study is performed with 2.5 USD per Mcf for the natural gas price and 4.5 USD per Mcf for the LNG price. By applying those prices, the economical ranges of each process are figured out. In addition, the natural gas liquefaction process selection strategies are suggested. There is a possibility that different prices of the natural gas and the LNG could show different economical ranges of various liquefaction processes. However, the optimizations with the profit optimization model which is developed in this thesis can be also applied to the process selection strategies for different situation. In addition, the specific cost and specific profit portfolios for various process designs by size can be applied to the supply chain network research field, and it can make the result more rigorous by multi-scale problem formation. Moreover, the specific profit portfolio approach can be applied to other chemical process industries.

REFERENCES

[1] **Mortazavi, A.,** *et al.*, "Enhancement of APCI cycle efficiency with absorption chillers," *Energy*, **35** (9), pp. 3877-3882 (2010).

[2] **Lee, I., et al.,** "Decision Making on Liquefaction Ratio for Minimizing Specific Energy in a LNG Pilot Plant," *Ind. Eng. Chem. Res.*, **54** (51), 12920-12927 (2015).

[3] **Lim, W.,** *et al.*, "Current status and perspectives of liquefied natural gas (LNG) plant design," *Ind. Eng. Chem. Res.*, **52** (9), 3065-3088 (2013).

[4] **U.S. Energy Information Administration,** "U.S. Natural Gas Prices," https://www.eia.gov/dnav/ng/ng_pri_sum_dcu_nus_m.htm, (accessed on Apr. 2017).

[5] **Lee, I., and Moon, I.,** "Economic Optimization of Dual Mixed Refrigerant Liquefied Natural Gas Plant Considering Natural Gas Extraction Rate," *Ind. Eng. Chem. Res.*, **56** (10), 2804-2814 (2017)

488800 Isolation of Low Grade Biobased Bulk Chemicals

Abstract:

In many applications present biorefinery suffers from the problem of limited utilization of products formed. Pulping of wood produces about or even less than 50% cellulose, while several byproducts need to be incinerated because of economic reasons. Typically, black liquor from pulping does contain appreciable amounts of carboxylic acids besides sugars and other byproducts. Targeting complete utilization of these low grade bulk chemicals does need severe technological intensification of downstream processing.

Focus of this project was the intensification of isolation of acetic acid and formic acid from black liquor evaporation condensates. This ternary mixture exhibits a binary high boiling azeotrope and a ternary saddle point azeotrope with the low boiling component water in high excess. Liquid-liquid extraction is well investigated and perfectly suited for separation of the carboxylic acids from the aqueous feed. However, solvent regeneration and byproduct isolation are still too energy intensive.

Combining distillation with an appropriate chemical reaction may intensify the solvent regeneration. Esterification of the carboxylic acids with a low molecular weight alcohol produces low boiling esters. These esters can be continuously removed from the reaction broth by distillation, acting positively on the equilibrium composition of the aqueous broth towards complete conversion of the target constituents. Expectedly catalytic acceleration of esterification would contribute to intensification. By using a reactive extractant with preference to acids the liquid-liquid equilibrium is improved compared to physical extraction and the acidic catalyst is kept in the solvent phase during extraction due to adduct formation with the reactive extractant. Admixture of solvent-soluble acidic coextractants may for sure limit the extraction capacity of the solvent, but provide the catalytic support. In solvent regeneration via reactive distillation the elevated temperature in the reboiler releases the catalyst from the adduct.

For technology assessment and process validation experiments were performed with Cyanex®923 (a Lewis basic phosphonate extractant) for extraction plus 4-dodecylbenzenesulfonic acid for catalysis. With this solvent mixture complete conversion and isolation of formic acid was achieved and methyl acetate yield of 88 % was obtained. For comparison distillative isolation of acids would cause an economic disaster and isolation by extraction/distillation would end up in the ternary azeotropic mixture.

498255 An Efficient Approach for Chemical Process Development Using Kinetic Modeling in Batch and Continuous Mode

Abstract:

Traditionally active pharmaceutical ingredients are manufactured in batch mode, however the pharma companies express increasing interest in continuous processing. Continuous manufacturing is an attractive alternative due to its benefits such as reduced development time, greater process safety when employing hazardous chemistries or extreme conditions, reduced operating costs, improved process control and product quality.

Kinetic modeling plays an important role in development stages of chemical processes regardless of the operation mode. Precise knowledge of the kinetic parameters is necessary for the scale up from laboratory to industrial scale and for the design of robust and safe processes.

This work describes an efficient approach for the development of a reaction in an API synthesis. The approach includes screening of the reaction conditions, estimation of the kinetic parameters both in batch and flow mode and their comparison, and also fine-tuning optimization applying Design of Experiments.

In the screening phase, we tested both conventional and microwave heating to find the most favorable conditions resulting in 99% conversion and we analyzed the possibilities for a continuous process. Specific sets of experiments were performed in both operation modes, kinetic curves and parameters were then obtained and assessed to determine if the studied variables impacted the impurity formation. The model was used to predict conversion values and impurity levels for different conditions. In the last stage optimization studies were carried out in order to determine the Design Space ensuring the product quality within specifications.

We demonstrate our methodology applied for a case study following the Quality by Design concept, including experimental data and also simulations. This approach can be used for the development of batch or continuous processes in the future. The use of kinetic modeling allowed deeper understanding of the reaction mechanism and its influential factors, resulting in shorter development time and enhanced confidence level regarding the impurity formation.

487968 Intensification of Carboxylic Acid Isolation Processes By Emulsification

Abstract:

Process intensification holds a central role in the biorefinery, as the substitution of crude oil by renewable feedstock overstrains standard processes in terms of economy and efficiency. Intensification can be achieved by applying heuristic methods to develop innovative processes in biomass refining. While conversion of biomass to any desired extend is the lesser problem (at least from the point view of downstream processing), the downstream processing of effluents is energy and resource demanding. Therefore, low grade carboxylic acids or low grade alcohols are rarely processed. Within isolation, low concentration and unfavorable combinations of separation relevant substance properties form the most troublesome problems. A promising concept to deal with this economic problem is the change of properties like solubility or relative volatility by chemical conversion.

The change of properties, a tool of process intensification, can be combined within existing processes and equipment. Esterification of low grade carboxylic acids for example does change physical properties significantly. For esterification of carboxylic acids (e.g. acetic acid or formic acid) high excess of water is a major obstacle. Superimposition of the chemical conversion with liquid-liquid extraction can be used to overcome this limitation. Enhancement of the reaction rate as well as mass transfer, may be attained by catalyst-induced emulsification. Surface active, acidic catalysts like sulfonic acids (e.g. *4*-dodecylbenzenesulfonic acid) accord with these requirements.

As a test system 60 g L^{-1} acetic acid was chosen with *n*-octanol as reactant and *n*-undecane as diluent. *4*-dodecylbenzenesulfonic acid was used as catalyst, due to its good performance in esterification catalysis and due to its emulsification properties. Experiments were performed in order to determine the influence of catalyst, reaction temperature and solvent composition on the reaction kinetics and separation efficiency. Experimental results for esterification of dilute acetic acid with *n*-octanol in the emulsion regime have shown the applicability of pseudo-homogeneous rate laws for modeling the reaction kinetics. This implies a quasi-homogeneous state as a result of emulsification. For implementation of this intensified process, a mixer-settler setup is suitable as it allows higher conversions and separation efficiencies through multi-stage operation. Phase separation can be achieved through centrifugation or electrical splitting.

487339 Sustainable Pig Iron Production Via Reduction of Mineral Iron Carbonate with Hydrogen

Abstract:

Innovative technologies are necessary to limit CO_2-emissions from energy-intensive iron and steel production from carbonate based ores. Iron carbonate reserves are found worldwide including reservoirs in the United States (e.g. Bisbee and Antler Mine in Arizona) and China. Pig iron is produced from iron ores in a multi-step process. This process includes roasting and sintering (Eq. 1) of iron carbonate minerals. Instead of roasting in air to produce hematite suitable for iron production via the blast furnace process, we suggest direct reduction of iron carbonate [1] with sustainably produced hydrogen (Eq. 2) as a 'green' alternative.

$FeCO_3 + 0.25\ O_2 = 0.5\ Fe_2O_3 + CO_2$ (1)

$FeCO_3 + H_2 = Fe + H_2O + CO_2$ (2)

Carbon dioxide emission (due to the thermal decomposition of iron carbonate according to Eq. 2) could even be diminished in a two-step process. In the first step, iron carbonate is directly reduced to iron and carbon dioxide. In the second step, the process gas is fed to a reactor for catalytic carbon dioxide hydrogenation to yield value-added carbonaceous products. In order to develop an environmentally benign synthesis route, an insight into reaction mechanism and kinetics is important. Therefore, thermogravimetric linear heating rate data was recorded and analyzed according to the ICTAC recommendations [2]. Iron formation according to Eq. 2 follows a nucleation and growth mechanism, verified by the prediction of the measured data with an Avrami-Erofeev A2 reaction model. Thermodynamic analysis revealed that the energy demand for iron production can partially be provided by the outlined process gas treatment by means of catalytic carbon dioxide hydrogenation to produce hydrocarbons.

1. Baldauf-Sommerbauer, G., Lux, S., and Siebenhofer, M. (2016) Sustainable iron production from mineral iron carbonate and hydrogen. *Green Chem.*, **18** (23), 6255–6265.

2. Vyazovkin, S., Burnham, A.K., Criado, J.M., et al. (2011) ICTAC Kinetics Committee recommendations for performing kinetic computations on thermal analysis data. *Thermochim. Acta*, **520** (1–2), 1–19.

495880 a Novel Selective Ion Separation Process Using Battery System

Abstract:

The increase of demand for rare metals and the decrease of the available ores make it interesting to develop separation processes for extracting metal ions from solutions such as brine lakes, seawater, and industrial waste water. In general, that solutions contain a mixture of salts and the concentration of valuable resource is very low, so it is required to develop a highly selective resource recovery technology. Distillation, adsorption and membrane processes are well-known separation technologies for ion separation from solutions, however, the selectivity, energy requirement, and operation costs are not sufficient. Therefore, a new approach to develop an efficient ion separation process is needed. In this work, we investigated the highly selective and energy efficient resource recovery process using battery system. The system consists of two battery electrodes which have ion selectivity. It operates a similar way to rechargeable batteries, and target ions are selectively extracted and released by charging/discharging step. Using this battery system, lithium or sodium ion are successfully separated from salt water even the solutions contain highly concentrated various ions (Yoon *et al.*, *Phys. Chem. Chem. Phys.*, 2013; Yoon *et al.*, *Energy Environ. Sci.*, 2014, Yoon *et al.*, *Chemosphere*, 2015). This battery system has not only energy saving process with efficient energy consumption but also eco-friendly process without chemical adding during the operation. The battery system could be a suitable method for ion separation process for selectively extracting metal ions.

Reference

Jaehan Lee, Seung-Ho Yu, Choonsoo Kim, Yung-Eun Sung, **Jeyong Yoon**, Highly selective lithium recovery from brine using a Î»-MnO_2-Ag battery, *Physical Chemistry Chemical Physics*, 2013, 15, 7690-7695.

Jaehan Lee, Seoni Kim, Choonsoo Kim, **Jeyong Yoon**, Hybrid capacitive deionization to enhance the desalination performance of capacitive techniques, *Energy & Environmental Science*, 2014, 7, 3683-3689.

Seoni Kim, Jaehan Lee, Jin Soo Kang, Kyusik Jo, Seonghwan Kim, Yung-Eun Sung, **Jeyong Yoon**, Lithium recovery from brine using a Î»-MnO_2/activated carbon hybrid supercapacitor system, *Chemosphere*, 2015, 125, 50-56.

Â

496638 Computational Investigation of Liquid Maldistribution in Periodically Operated Structured Packed Beds

Abstract:

The non-linear hydrodynamics combined with multiscale transport processes engenders the modeling of packed bed reactors (PBR) a challenging task, particularly in addressing the scale up issues for its commercial application. Familiarized with the steady state hydrodynamics over the past two decades, researchers are now extensively exploring the efficacy of unsteady state operation of packed bed reactors in process intensification and prolong reactor service life. Despite the proven advantages, implementation of periodic flow operation in commercial PBRs is still far from reality, mainly due to process control safety. Although, the application of steady state PBR for commercial purposes is well established, the design concept of periodic PBR needs more attention, even heuristically, to understand the influence of all key parameters. With enhanced capabilities of CFD methods, computational studies can not only complement various facets of physical phenomena but also can help in understanding certain aspects that are generally not attainable by experiments. Surprisingly, limited researches have attempted to develop CFD based models for unsteady PBRs to understand the influence of operating parameters. Moreover, none of these studies have addressed the effect of packing arrangements on flow distribution during periodic operation. Additionally, hydrodynamics of periodic PBRs, fed with non-Newtonian liquids, has rarely been addressed. Cyclic or periodic operation, dictates the periodic toggling of a fluid phase at the inlet, between a low-level (base) and a high-level (peak), typically referred to as the min-max flow operation. This further is termed as the on-off flow when the base flow rate is set to zero.

In this article, we aim to develop a 3-D CFD model for periodic PBRs to understand the effect of particle arrangements and various splits under min-max and on-off operation on flow distribution. A CFD model is developed based on unit cell approach incorporating single phase periodic flow to understand the intrinsic behavior of flow modulation in structured packings. To elucidate the effect of non-Newtonian liquid, a basic power-law model is employed in our validated CFD model. To examine the effect of particle arrangements, face centered cubic (FCC) and modified simple cubic (SC) packing orientations are investigated as shown in Fig. 1. Figure 1: Schematic showing grids and boundary conditions of (a) FCC, and (b) modified SC arrangements. Moreover, three different combinations of min-max and on-off operation are chosen under fast mode to identify the influence of flow splits. Liquid distribution is quantitatively examined at the mid-plane of the geometries by dividing the plane into twelve area segments and then evaluating the mass flux through each of the segments. An improved distribution is observed for periodic flow as compared to continuous operation, which is even more pronounced for min-max operation. A distribution coefficient, calculated using the standard deviation concept, indicates that the FCC packing shows the best distribution for min-max operation with the split having maximum peak flow duration as shown in Fig. 2. However, exactly opposite trend is observed in modified SC, where the split with the lowest peak flow rate duration shows the best distribution among the three splits which was even more pronounced for the min-max operation. FCC exhibits better liquid distribution than the modified SC although having a higher pressure drop value due to its lower porosity. Consequently, a comparitive analysis was conducted to determine the desired mode of periodic operation with optimum split ratio that results in least maldistribution. This study will not only find relevance to several packed bed applications but also indicate the periodic flow operation as a viable process intensification technique.

Figure 2: Mass flux distribution at the mid-plane of FCC for different n values at t=10 s for (a) n=0.6, (b) n=1.0, and (c) n=1.2.

490465 Overcoming Challenges for Scale-up of Continuous Pharmaceutical Drug Substance Processes

Abstract:

The area of continuous technology and reactor design applied to pharmaceutical drug substance processes has advanced significantly over the last decade. Primary drivers of flow versus batch operations are enhanced safety, greater process control, increased productivity, and better operational stability. However, several obstacles need to be overcome prior to successfully scaling-up flow processes from laboratory to manufacturing. Some examples are solid fouling, stability process hold, and mapping feasible operational space for process perturbations. This talk will focus on two case studies related to a: 1) Grignard reaction carried out in a continuous stirred tank reactor (CSTR) and, 2) hydrogenation reaction run in a pipes-in-series plug flow reactor (PFR). The investigated case studies will illustrate methodology and use of experimental tools in conjunction with mathematical models to overcome the above challenges. Specifically, the Grignard case study involved issues with instable process holds and solid fouling issues (affecting downstream performance) that were overcome by mechanistically probing the system with careful experimental design and analytical techniques. On the other hand, mathematical models were used for the hydrogenation case study to demonstrate their power in mapping out the feasible operational space. These efforts enabled scaling up these reactions to successfully run in various pivotal and validation campaigns.

485306 Preparation of Hazardous "on-Demand" Reagents Using Continuous Processing

Abstract:

Hazardous and unstable reagents are often forgone for use during manufacturing due to storage and safety concerns. These reagent are often replaced with more expensive alternatives. Continuous processing allows for the preparation of hazardous and/or unstable reagents that can be used directly in the production of pharmaceuticals. This talk will focus on using the Corning ® Advanced Flow Reactor™ (AFR) for the continuous preparation and use of several hazardous and unstable reagents.

489921 Model-Aided Development for a Continuous Amide Bond Formation in a Drug Substance Manufacturing Process

Abstract:

Mathematical models are valuable tools in development to determine acceptable parameter ranges and control strategies for continuous manufacturing. Some of the benefits of modeling in pharmaceutical development are its ability to reduce material costs and resource needs, while simultaneously increasing process understanding and quality control.

A case study is presented here that demonstrates the use of NMR reaction monitoring (ReactNMR), to gather high-value reaction data in batch mode for kinetic parameter estimation. The model describes the amide bond formation between an aniline and a mixed anhydride. The kinetic model was then employed to describe a continuous process that includes a plug flow reactor (PFR), a delay coil, two mixed-suspension, mixed-product-removal (MSMPR) crystallizers in series, and a filter/dissolution vessel. The resulting model was used to enable risk assessment, test failure modes, and build decision trees for use in the event of a process upset at manufacturing scale.

492008 Effects of Catalyst and Metals Pretreatment for the Continuous Hydrogenation of a Halogenated Aromatic Nitro Compound

Abstract:

Continuous processing for the synthesis of active pharmaceutical ingredients (APIs) continues to be a growing area of interest and investment. Continuous heterogeneous catalysis offers both safety and capital cost advantages over a traditional batch autoclave.

This investigation explored impurity control for the continuous hydrogenation of a halogenated aromatic nitro compound. The hydrogenation utilized a palladium catalyst for the reduction of an aromatic nitro compound to an aniline. In particular, the work involved development of a packed bed reactor for the continuous heterogeneous catalysis. Several challenges appeared in transitioning the chemistry from batch to continuous processing. These challenges were characterized using advanced monitoring techniques (flow-NMR, online HPLC, etc.) and investigated using a combination of batch and flow processing. As part of that effort, a catalyst pretreatment was investigated which had been demonstrated as effective in batch mode for suppressing an impurity. Parameters investigated included duration and temperature of pretreatment, as well as liquid flow profile. Also investigated was the effect of residual metals in the feed solution entering from a previous synthetic step. Feed pretreatment and its impact on reaction impurity profile and bed life were evaluated.

492261 Large Scale Production of Pharmaceutical Intermediates in the Flow Mode

Abstract:

In recent years, there have been significant efforts in the Pharmaceutical Industry to adopt continuous processing technologies to support the chemical/unit operation steps. Major drivers for developing flow processes and continuous manufacturing for specific steps include process safety issues, poor heat transfer, poor mass transfer in multiple phase reactions, presence of unstable intermediates, extreme reaction conditions requirements, and the need for short residence time. Since 2015, STA Pharmaceutical, a WuXi AppTec company, has established a leading continuous processing technology platform to support process R&D as well as plant scale production of pharmaceutical intermediates. The reaction types supported to date include cryogenic and room temperature lithiations, deprotection, photochemistry, ozonolysis, oxidations, dehydrations, coupling, cyclization, amination, and azide chemistries among other reactions. Flow steps included the above chemical steps as well as unit operations including phase separation, crystallization, filtration, and dissolution. The scale of the operations ranged from a few hundred grams to a few hundred kilograms under GMP and non-GMP modes. This presentation gives an overview of the continuous processing platform in STA followed by case studies illustrating the key benefits of this technology platform.

494952 Plasmonic Metal Nanocatalysts As Platform for Continuous Synthesis of Drug Substances

Abstract:

Plasmonic Metal Nanocatalysts as Platform for Continuous Synthesis of Drug Substances

Farshid Mohammadparast, Andishaeh Dadgar, Marimuthu Andiappan, School of Chemical Engineering, Oklahoma State University, Stillwater, OK, USA

Approximately 25% of all the reactions performed in the pharmaceutical industry consist of cross-coupling reactions. The cross-couplings have been traditionally carried out by homogeneous palladium (Pd) catalyzed batch processes. The National Science and Technology Council (NSTC) has recently identified the continuous manufacturing of pharmaceuticals as one of five manufacturing areas of emerging priority in the USA, revealing that continuous manufacturing of pharmaceuticals has the potential to reduce manufacturing cost by up to 40-50%. Given these advantages, in recent years, metal nanoparticle catalyzed cross-couplings have been of great interest since they can be used as a platform to switch cross-couplings from traditionally used batch mode to continuous flow mode. Additionally, heterogeneous metal nanocatalysts can be immobilized on a rigid support in the form of a packed-bed reactor, and therefore, can avoid the need for subsequent difficult steps to separate the catalyst and product.

In this contribution, we show that heterogeneous plasmonic metal nanocatalysts can be used as efficient catalysts to drive cross couplings. Specifically, we demonstrate that visible light can be used as a powerful tool to make plasmonic metal nanocatalysts as efficient photo-catalysts with activity and stability superior to that of traditionally used homogeneous Pd based catalysts. These plasmonic metal nanocatalysts can be used to design efficient cross-coupling flow processes to produce drug substances.

Abstract, AIChE 2017

Developing Scale-Up Approach for Fast Chemical Reactions in Continuous Flow

Because of potential high API demand for a particular Merck product currently under development, a new chemical synthesis was developed with improved yield. One of the chemical transformations in the synthesis was developed as a continuous flow step, improving reaction conversion by over 15% compared from batch mode. The reaction exhibits very fast kinetics (in the order of ms) requiring very good mixing. Successful scaling up of such fast reactions requires two important pieces of information:

1. Understanding of how reaction performance is affected by mixing.

2. A reliable way of measuring/scaling up mixing across different scales/geometries of static mixers.

In order to gain understanding on the effect of mixing, the continuous reaction was ran at various flow rates, static mixer geometries/sizes and temperatures. The collected data for reaction conversion was analyzed statistically in order to obtain a mathematical model for its dependence on the above process parameters. The model had acceptable predictive capabilities and confirmed an expected trend: for each mixer, the conversion was increased with increasing flow rate (improved mixing) and decreasing temperature (slower kinetics, allowing for better mixing). The drawback of this approach was that all these models were static mixer specific. If a different static mixer was to be used, a new study was needed in order link the flow rates to reaction conversion. A different, more generalized method was needed to measure/characterize mixing, independent of the mixer type.

Many research attempts to characterize and scale-up continuous mixing in static mixers usually involve correlations using scale-up parameters such as Reynolds number (Re), residence time (τ) or turbulent energy rate (ε). To link all these scale up parameters to the mixing intensity, a method known as the Fourth Bourne Reaction is usually employed. Some of these parameters usually work well, however they may fail to capture any differences in the internal geometry of the static mixer. Instead of using the above scale-up parameters, the conversion from the Fourth Bourne Reaction was used directly as a surrogate process parameter. This direct method of measuring mixing is more accurate as it will capture any internal changes to the mixing pattern (such as back mixing), which may not be captured if the other scale-up approaches are used. It can be seen as a snapshot of the mixing pattern inside the static mixer.

In order to show the applicability of this method, the Fourth Bourne Reaction tests were performed on the same mixers used in the continuous reaction, within the same flow rate ranges. For each mixer type/flow rate combination, the Bourne reaction conversion (BRC) number was assigned and used as a surrogate process parameter. Then the BRC and temperature only were used in the same statistical analysis. Both BRC and temperature were found to be statistically significant and the overall model had the same predictive capabilities. The advantage of this method was the ability to substitute the mixer type/flow rate combinations with the single BRC parameter. This allowed the three model plots produced previously for each of the mixers to be reduced to only one: BRC and temperature only. This methodology also makes scale-up much more straight-forward: one need to show only that the BRC is the same across scales in order to ensure identical mixing (provided the BRC test is executed identically).

497306 Economic and Energetic Analysis of Biofuel Supply Chains

Abstract:

Economic and Energetic Analysis of Biofuel Supply Chains

Rex T. L. Ng[1,2] and Christos T. Maravelias[1,2]

[1]Department of Chemical and Biological Engineering and [2]DOE Great Lakes Bioenergy Research Center, University of Wisconsin-Madison, Madison, WI

The efficient and effective transportation of lignocellulosic biomass is one of the critical challenges in biofuel production, due to its low bulk density and durability properties. Generally, in a biofuel supply chain (SC), baled biomass is either transported to a biorefinery directly, or sent to a depot for densification prior to transporting densified biomass to a biorefinery [1]. Many studies show the installation of depots in biofuel SCs offer better economic performance in large scale biorefineries [2–4]. However, from an energy point of view, depots may be infeasible if the total primary energy input (e.g., biomass harvesting and collection, depot operation, and baled and densified biomass transportation) is higher than the energy content of biofuel.

To better understand what supply chain configurations are preferable under different scenarios, we develop systematic methods to compare the cost and energy efficiency of biofuel SC under different configurations. We define a configuration as a combination of depot selection (with and without depots) and transportation modes (truck and rail). The analysis aims to provide some insights into the following questions: *When should we ship biomass directly to the biorefinery? When depot is economically and energetically feasible? Which transportation mode yields low cost and high energy efficiency? What is the trade-off between the improvement of densification technology and cost or energy efficiency?*

In this analysis, we consider corn stover which is collected from 12 Midwest states in the US. We assume that corn stover productivity is on an annual basis. A high moisture pelleting process [5] is employed at the depot, while dilute acid pretreatment, separate hydrolysis and fermentation are employed at the biochemical biorefinery [6]. Note that the analysis can be repeated using the proposed method for any feedstock, technology, and type of productivity (annual, seasonal or monthly basis). We estimate the minimum ethanol selling price (MESP) and total energy input based assuming the feedstock collection area to be circular with the biorefinery or the depot located at its center. When a combination of two or more configurations leads to better economic or energy benefits (referred to in this work as 'hybrid configuration'), the average distance from depots in the annular area to the biorefinery at the center is estimated.

The results show what reductions in cost and energy input can be achieved through a hybrid configuration (Fig. 1A). From an economic point of view, direct shipment of baled corn stover (orange region) is preferred if the distance between biorefinery and harvesting site is lower than 155 km. Depots installation and truck transportation of densified corn stover (blue region) are favorable if the distance ranges between 155 to 260 km; while rail transportation of densified corn stover (red region) will only lead to better economic benefits at larger distance (=260 km). From an energy point of view, rail transportation of densified biomass is always better than truck transportation. The depot installations with rail transportation have lower total energy input values than those with truck shipment of baled biomass if the distance is greater than 170 km. We further discuss key cost and energy efficiency drivers: corn stover productivity, biorefinery size, and densification efficiency. The MESP decreases with the increase of productivity and biorefinery size (Fig. 1B). We will show how the improvements in densification technologies can reduce transportation cost and energy consumption.

Finally, we will briefly discuss how the finding of this study can be used to formulate tractable mixed-integer programming (MIP) models for biofuel SCs operational planning, which account for non-uniform biomass availability and other practical considerations. We will show results from two case studies in southern Wisconsin.

Figure 1: (A) Hybrid configuration, and (B) MESP as a function of biomass productivity and biorefinery capacity.

References:

[1] R.T.L. Ng, C.T. Maravelias, Design of biofuel supply chains with variable regional depot and biorefinery locations, Renew. Energy. 100 (2017) 90–102.

[2] D.J. Muth, M.H. Langholtz, E.C.D. Tan, J.J. Jacobson, A. Schwab, M.M. Wu, et al., Investigation of thermochemical biorefinery sizing and environmental sustainability impacts for conventional supply system and distributed pre-processing supply system designs, Biofuels, Bioprod. Biorefining. 8 (2014) 545–567.

[3] S. Kim, B.E. Dale, Comparing alternative cellulosic biomass biorefining systems: Centralized versus distributed processing systems, Biomass and Bioenergy. 74 (2015) 135–147.

[4] P. Lamers, E.C.D. Tan, E.M. Searcy, C.J. Scarlata, K.G. Cafferty, J.J. Jacobson, Strategic supply system design – a holistic evaluation of operational and production cost for a biorefinery supply chain, Biofuels, Bioprod. Biorefining. 9 (2015) 648–660.

[5] J.S. Tumuluru, K.G. Cafferty, K.L. Kenney, Techno-economic analysis of conventional, high moisture pelletization and briquetting process, in: Am. Soc. Agric. Biol. Eng. Annu. Int. Meet. 2014, ASABE 2014, 2014: pp. 4177–4189.

[6] D. Humbird, R. Davis, L. Tao, C. Kinchin, D. Hsu, A. Aden, et al., Process Design and Economics for Biochemical Conversion of Lignocellulosic Biomass to Ethanol: Dilute-Acid Pretreatment and Enzymatic Hydrolysis of Corn Stover, Natl. Renew. Energy Lab. (NREL), Golden, CO. (2011).

489815 Synthesis of Sustainable Processing Networks: Location-Dependent Biorefinery Models

Abstract:

The current and projected increase in demand of commodities, water and energy caused by the growing world population motivates innovation in process synthesis-design towards more sustainable processes. This is achieved by using alternative (renewable) raw materials, such as biomass, incorporating new process technologies and satisfying new design objectives and constraints, including sustainability.

This work describes the development, modeling and optimization of a comprehensive biorefinery superstructure with multiple product and feedstocks. Over 100 processing alternatives for converting the mentioned feedstocks into the various products are included, leading to over a hundred trillion theoretical alternative solutions.

A systematic process synthesis framework for biorefineries has been developed (Bertran et al, 2017), alongside the necessary methods and tools (such as a comprehensive database of biomass, chemicals, reaction paths, technologies, prices and many more), and implemented in a software interface named Super-O. This framework has been applied in this contribution to solve various problems derived from the biorefinery superstructure, under different scenarios. These include: (i) selection of a product or a set of products given the feedstock mix; (ii) selection of a feedstock or set of feedstocks given the product mix; (iii) selection of the processing route for a given feedstock-product combination; (iv) selection of the location given the feedstock, product and route; (v) all possible combinations of the aforementioned options. Moreover, multiple location options are investigated, by incorporating transportation options between each location, to determine the optimal distributed location-configuration for the entire value chain. Using integer cuts, a number of top-ranked solutions are generated and analyzed.

In this contribution, a strong emphasis is made on the management of the problem data through a data structure implemented as a database. Upon a specific problem definition, the relevant data is retrieved from the database and the problem is then solved via optimization.

The production of bioethanol alongside co-products (such as succinic acid, lactic acid, and lysine) from various lignocellulosic feedstocks is used as a case study to determine the optimal feedstock mix, processing route, set of locations and configuration.

References:

- M.-O. Bertran, R. Frauzem, A.-S. Sanchez-Arcilla, L. Zhang, R. Gani, Comput. Chem. Eng. (2017) DOI: 10.1016/j.compchemeng.2017.01.030.

500097 Electrochemistry As a Sustainable Alternative for Distributed Processing of Biomass

Abstract:

Electrochemical processes use electrons, a green reagent, as main driving force for the conversion of chemical energy into electrical energy of vice versa. Conventional electrochemical technologies (i.e. fuel cells and batteries) are well developed and are ideal for distributed energy generation and storage. Recent developments have extended scope to include stabilization and energy densification of biomass feedstock to produce transportation fuels. Among the main advantage of electrochemical technologies for distributed processes are that it can features rapid start up and turn off, which enable them to operate either in batch or continues operations and be directly fed by intermittent renewable energy sources, such as solar and wind. In the specific case of distributed biomass processing, electrochemical processes can be implemented the conversion of generated CO_2, upgrading of lignocellulose biomass, and biomass stabilization.

In this work the applicability of electrochemical technologies on distributed biomass processing schemes will be contextualized. Current state of the art and potential impacts will be assessed based on technology readiness, roadblocks, and viability. Case studies will be presented for the electrochemical reduction of CO_2, and the electrochemical upgrading/stabilization of biomass derived streams.

499969 Assessing Fuel and Feedstock Energy Use in the U.S. Chemical Sector: A Supply Chain Analysis

Abstract:

Assessing Fuel and Feedstock Energy Use in the U.S. Chemical Sector: A Supply Chain Analysis Scott Nicholson Alberta Carpenter Rebecca Hanes

In chemical manufacturing, fossil fuels are consumed both as an energy source and as feedstocks that are converted to higher-value chemicals. This work expands on the U.S. Department of Energy's chemical manufacturing bandwidth study, which quantified the potential for reducing direct energy consumption in the chemicals sector through the adoption of the most energy efficient manufacturing technologies available [1]. The scope of the bandwidth study was limited to fuels consumed within the plant boundary and fuels consumed for electricity generation. Fuel consumed upstream in the supply chain and fuel consumed as feedstock (defined as fuels that are converted to chemical intermediates rather than being used as an energy source) were not quantified; by one estimate, 60% of the fossil fuels consumed in the U.S. chemical manufacturing sector are consumed as feedstocks [2]. The present study quantifies, in addition to process energy, energy consumed upstream in the supply chain and energy embedded in fossil feedstocks and identifies chemical supply chains in which an understanding of the supply chain can help to identify opportunities for reducing total fossil fuel consumption. Alternative biomass-based processes, which require no fossil feedstocks but do not necessarily consume less energy, are then analyzed to determine if the alternative processes can reduce the total fossil fuel consumption in the identified supply chains. Reducing the chemical industry's reliance on fossil fuels could lead to greater long-term profitability in the industry, as crude oil prices are projected to rise steadily over the coming decades [3].

The Materials Flows through Industry (MFI) supply chain modeling tool, developed at the National Renewable Energy Laboratory, is employed to quantify supply chain energy and feedstock requirements for the chemicals analyzed previously in the bandwidth study. The MFI tool is a flexible framework for modeling material and energy usage in the supply chains of commodities such as chemicals, primary metals, and paper products [4]. MFI contains U.S.-specific data on energy and material inputs to 1,469 manufacturing processes for 707 commodities. U.S. commodities produced via multiple technologies are modeled using weighted averages of the current production technology mix based on market shares reported in the literature [5]. For this study, energy values are calculated at both the immediate process level and as a cumulative total for 10 steps up a chemical's supply chain. In MFI, a "step" consists of the material and energy inputs to the intermediate products manufactured in the previous step, which then become the products of the next step in the supply chain. Material and energy inputs beyond the 10th step have been shown to be negligible compared to the first 10 steps, so the supply chain calculations are truncated at that step.

MFI is similar to standard life cycle assessment (LCA) software packages, but the tool goes beyond these platforms in several ways. First, MFI is adaptable; processes and process weightings can be adjusted individually to model existing or future manufacturing scenarios. Modeling alternative production technologies in a particular supply chain is therefore straightforward. MFI also has the capability to disaggregate supply chain energy requirements into three types: process fuel, energy in fuels used as feedstocks, and fuel for electricity generation. For each chemical in the current analysis, the MFI model is used to calculate process and feedstock energy on a per-kilogram basis. These per-kilogram values are multiplied by annual U.S. production quantities from 2014 or later [5] to determine the energy requirements needed to produce the given chemical at the national scale. These calculations are repeated for alternative and biomass-based pathways for chemicals with feedstock-intensive supply chains, in order to compare the total fossil fuel consumption in the conventional and alternative supply chains. When comparing alternative and conventional petroleum production routes, the former may use significantly less feedstock energy, particularly for biomass-based production routes. However, alternative production routes may or may not use less process energy; in the case of advanced, non-commercial technologies, process energy is likely to increase and this increase may outweigh any savings in feedstock consumption. Both feedstock consumption and energy use must be quantified to determine which, if any, alternative technologies offer a supply-chain-level reduction in total fossil fuel consumption over the conventional technologies.

One of the chemicals in the current analysis is 1,3-butadiene, a chemical intermediate used predominantly in the production of synthetic rubber [6]. Essentially all commercial-scale production of butadiene in the U.S. derives from the byproduct streams of ethylene cracking [7]. Since ethylene is derived from fossil feedstocks, there are significant feedstock-embedded energy inputs in the conventional butadiene supply chain. Preliminary results show that supply chain feedstock consumption for 1,3-butadiene is 220 GJ/tonne, which exceeds both supply chain scale process energy consumption (21 GJ/tonne) and the onsite energy consumption reported by the bandwidth study (18 GJ/tonne) by a full order of magnitude. The alternative manufacturing processes for 1,3-butadiene are all biomass-based and begin with the fermentation of sugars [8]. These processes, because they do not involve ethylene cracking, are expected to reduce feedstock consumption in the 1,3-butadiene supply chain significantly. However, these processes have yet to be optimized for commercial scale production, and may show an increase in both onsite and supply chain energy consumption. The other chemicals analyzed in this study are caprolactam, para-xylene, and propylene glycol, all of which have biomass-based alternative processes under development [8].

This analysis quantifies the technological potential of alternative chemical manufacturing processes to reduce fossil fuel consumption. The economic implications of transitioning chemical manufacturing to next-generation energy-efficient and/or biomass-based production technologies are outside the scope of this study, as are the logistical complexities involved in shifting from a fossil-based to a biomass-based chemical sector. Still, the results of this study could be used to inform decisions around the commercialization of advanced manufacturing processes. They can be used to determine whether it is more worthwhile, from the perspective of reducing fossil fuel consumption, to upgrade an existing manufacturing process or to move to another technology entirely. Future extensions of this work may explicitly analyze the process economics of these alternative production technologies by incorporating costs of material and energy inputs. If these results show that the bio-based technologies can both reduce feedstock consumption *and* be economically competitive, then a stronger argument can be made for commercialization in the near future. In contrast, an alternative process that is significantly more expensive would still require further development. Lastly, the MFI tool accounts for energy use from resource extraction through production (cradle-to-gate); it does not currently look at use-phase or end-of-life energy requirements (i.e. the full cradle-to-grave). These latter stages are beyond the scope of this chemical sector analysis, since the downstream products manufactured from these commodity chemicals are used across many sectors —not just the chemical sector. With additional data, MFI has the ability to capture use-phase and end-of-life, and this is another potential area of future research.

References

[1] Brueske, S., Kramer, C., & Fisher, A. (2015). *Bandwidth Study on Energy Use and Potential Energy Saving Opportunities in US Chemical Manufacturing* (No. DOE/EE-1229). Energetics. Retrieved from https://energy.gov/sites/prod/files/2015/08/f26/chemical_bandwidth_report.pdf

[2] U.S. Department of Energy, Energy Information Administration. Office of Energy Consumption and Efficiency Statistics. *2010 Manufacturing Energy Consumption Survey*. https://www.eia.gov/consumption/manufacturing/data/2010/

[3] U.S. Department of Energy, Energy Information Administration. *Annual Energy Outlook 2017 with projections to 2050* (No. DOE/EIA- 0383(2017)). Retrieved from http://www.eia.gov/forecasts/ aeo/pdf/0383(2017).pdf.

[4] Hanes, R.J. & Carpenter, A. (2017) Evaluating opportunities to improve material and energy impacts in commodity supply chains. *Environ Syst Decis 37: 6*. doi: 10.1007/s10669-016-9622-5

[5] IHS (2017). *Chemical Economics Handbook (CEH)*. https://www.ihs.com/products/chemical-economics-handbooks.html.

[6] White, W. C. (2007). Butadiene production process overview. *Chemico-biological interactions, 166(1)*, 10-14. doi: 10.1016/j.cbi.2007.01.009

[7] Sriram, P., Hyde, B., & Smith, K. (2016). Butadiene. *Chemical Economics Handbook*. IHS.

[8] Biddy, M. J., Scarlata, C., & Kinchin, C. (2016) *Chemicals from Biomass: A Market Assessment of Bioproducts with Near-Term Potential*. National Renewable Energy Laboratory (NREL). Retrieved from http://www.nrel.gov/docs/fy16osti/65509.pdf

501692 Integration of Fast Pyrolysis and Electrolyzer for Deoxygenation of Biomass

Abstract:

Domestically, the US is challenged with improving the environment, energy security, and energy productivity, thus many efforts are concerted in utilizing biomass as a viable source of renewable liquid transportation fuels and contributing to a robust carbon-neutral energy production. However, achieving a high carbon, hydrogen and overall energy efficiency is an ongoing challenge for current biomass upgrading technologies. The main approach of upgrading biomass to hydrocarbon fuels is by hydroprocessing, which relies heavily on large hydrogen consumption and on large, more centralized production scales. In this work, we present a different approach of using distributed pyrolysis unit scale with low hydrogen requirement by utilizing electricity to drive the deoxygenation of molecules present in the biomass. Currently, we have demonstrated the feasibility of electrochemical deoxygenation using solid oxide electrolysis cell in small cells, multi-cell stacks, and in an integrated test with a bench-scale fast pyrolysis unit. The electrolysis process removes oxygen from the oxygenated organic molecule as well as from steam to produce hydrogen in-situ allowing for a distributed, small scale integrated upgrading unit. Specifically, we will report recent work with a fast pyrolysis unit with 1 kg/h ponderosa pine feed, 480 deg C fast pyrolysis temperature and 550-600 deg C process at the electrolyzer. These process conditions allowed us to achieve an improved quality of the bio-oil. In this presentation, we will also discuss the present challenges of electrolyzer integration with a fast pyrolysis unit as well as the characterization of products using ^{13}C NMR, ultimate analysis (CHNOS), and GC-MS.

CLEAN ENERGY SMART MANUFACTURING INNOVATION INSTITUTE

Clean Energy Smart Manufacturing Innovation Institute (CESMII) Overview

Accelerating the Smart Manufacturing Transformation

Jim Wetzel

ManufacturingUSA®

An interagency team building partnerships with U.S. Industry and Academia

USDA

Manufacturing USA (14)

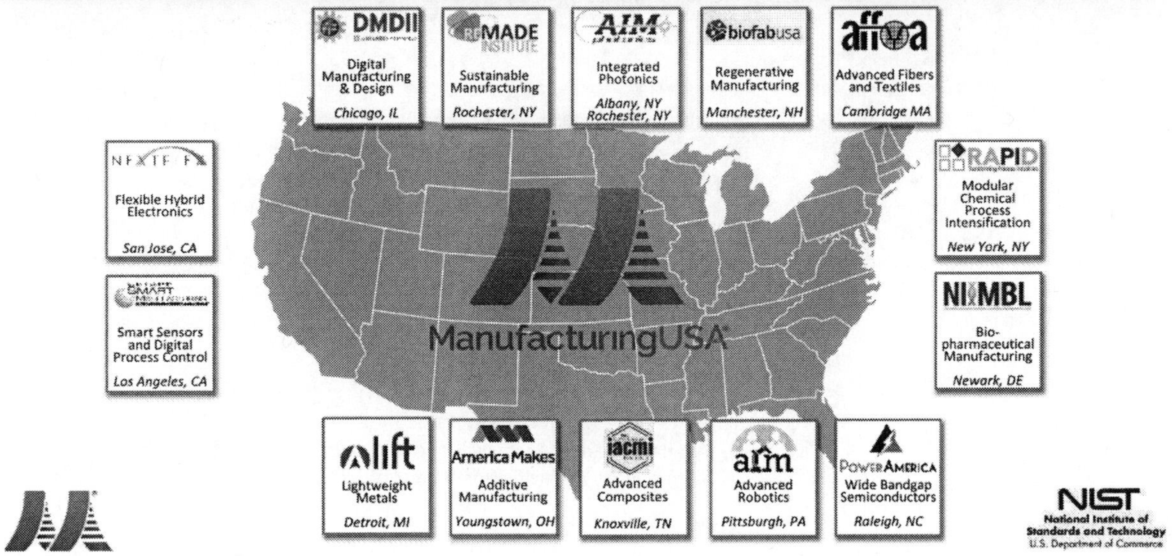

- DOE Selected - June 2016
- DOE Cooperative Agreement - Dec 2016
- Opened Institute – Jan 2017
- Headquarters – LA (UCLA)

ManufacturingUSA

CLEAN ENERGY SMART MANUFACTURING INNOVATION INSTITUTE

CESMII Reach is Comprehensive

A Network of Networks

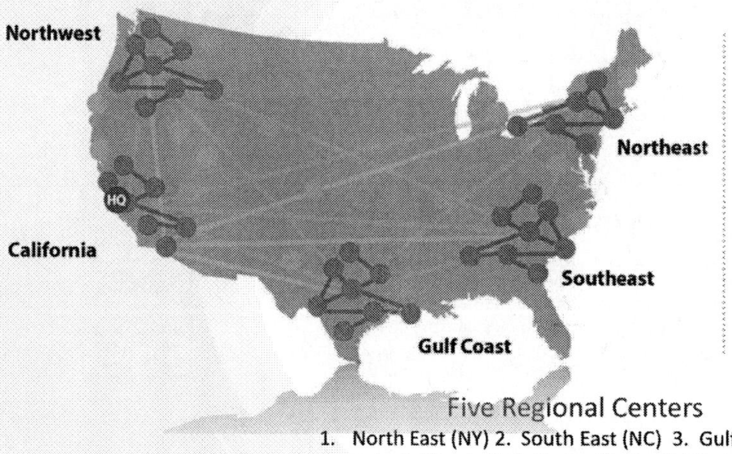

CESMII HQ

- Based in LA
- Institute Administration
- Central Platform and Marketplace Development and Management

Five Regional Centers

1. North East (NY) 2. South East (NC) 3. Gulf Coast (TX) 4. California (CA) 5. Pacific North West (WA)

CLEAN ENERGY SMART MANUFACTURING INSTITUTE

CESMII Mission – by the numbers

- Double energy productivity in US manufacturing every 10 years
- Halve the cost of deploying SM systems relative to state of the art in 5 years
- Increase the SM workforce in US multi-fold in 10 years
- Double the SM supply chain rate of increase in value and participation
- Reduce U.S. energy use in 10 years while increasing manufacturing competitiveness

CLEAN ENERGY SMART MANUFACTURING INNOVATION INSTITUTE™

What is Smart Manufacturing?

Smart Manufacturing (SM) is the business, technology, infrastructure, and workforce practice of optimizing manufacturing through the use of engineered systems that integrate operational technologies and information technologies (OT/IT).*

*Throughout the roadmap, these integrated engineered OT/IT systems are referred to as "cyber-physical systems."

CLEAN ENERGY SMART MANUFACTURING INNOVATION INSTITUTE

Smart Manufacturing: Converging OT and IT

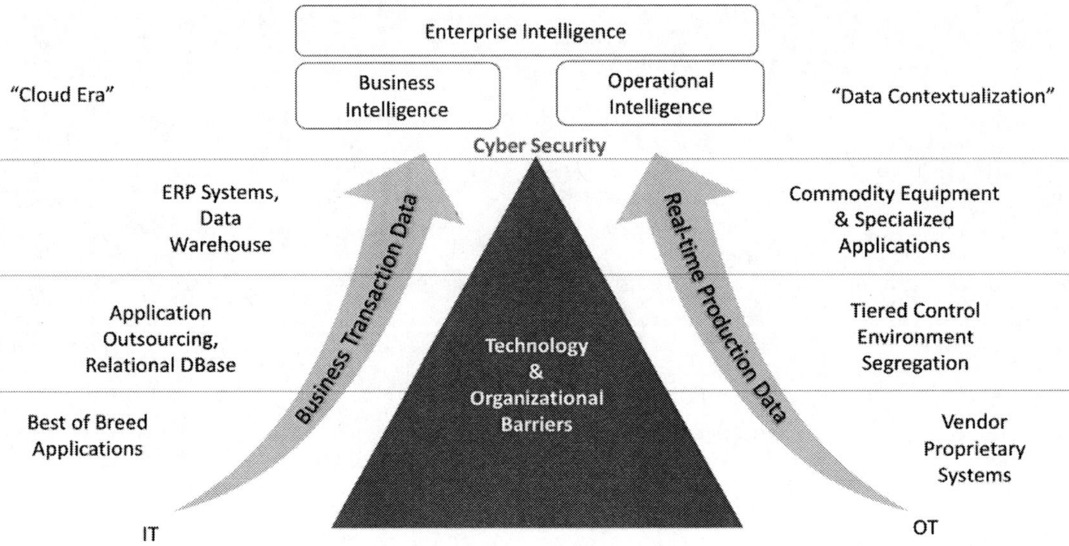

9

SM enables the right information and right technology to be available at the right time and in the right form to the right people, powering smart decision-making within factories and across entire value chains.

Small...Medium...Large Enterprises

CLEAN ENERGY SMART MANUFACTURING INNOVATION INSTITUTE

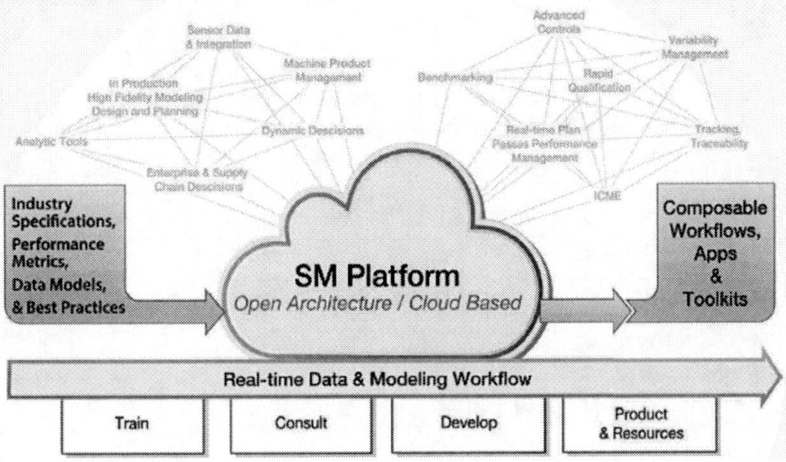

CLEAN ENERGY SMART MANUFACTURING INNOVATION INSTITUTE

A Glimpse of What we are WORKING on...

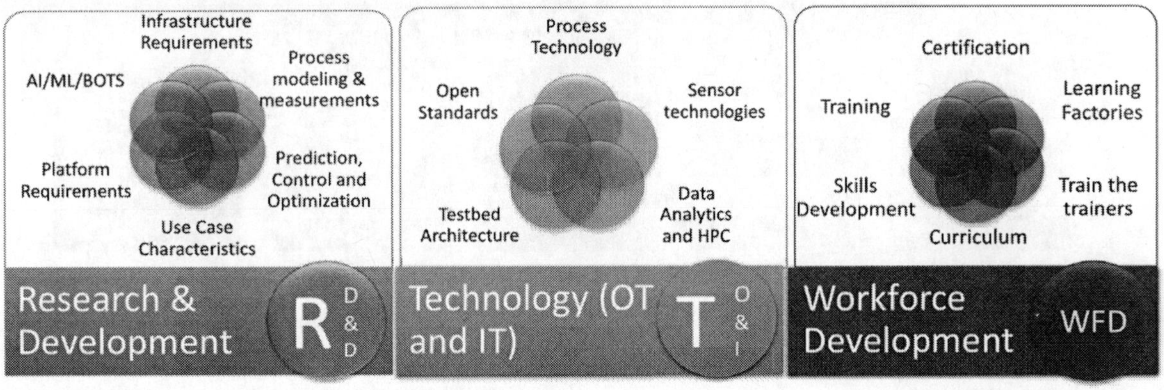

CLEAN ENERGY SMART MANUFACTURING INNOVATION INSTITUTE

So…How's your STITCHING?

It's all about the flow from beginning to end… and the connected threads

Become a Member

Strength of CESMII's Collaboration Network

Large	Medium	Small	Academia & Research, Govt. Labs, Associations	Vocational & Community Colleges, K-12
Industry: Users, Providers, Third-party Integrators				

- Investment in key industry-driven priorities
- Accelerate technology adoption
- Realize business results, energy sustainability and innovation
- Shared participation, non-proprietary results, IP
- Form relevant partnerships
- Solve collective challenges, catalyze R&D Investment
- Develop a strong workforce via education/training
- Improve economic development, increase jobs

CLEAN ENERGY SMART MANUFACTURING

14

We Meet You Where You Are

"Smart Manufacturing enables an ecosystem that makes the impossible, possible."

Bringing together the Ecosystem

- Large Companies (500+ employees/members)
- Medium Companies (>100 and <500 employees/members)
- Small Companies (<100 employees/members)
- Community Colleges, K-12 Educational
- Academics, Govt. Labs, and Non-profits

CESMII is tackling barriers that no one entity can address alone.

Advanced Technology
- Sensors
- Controls
- Platforms
- Models

Workforce
- Curriculum Development
- Certifications
- Training
- Nationwide strategy

Impact
- Energy
- Productivity
- Optimization
- Performance
- Accelerated technology adoption
- New business models

Economic Development
- Improved jobs
- Energy sustainability
- Competitiveness
- Innovation
- Revitalized communities and access

CLEAN ENERGY SMART MANUFACTURING

15

THANK YOU & QUESTIONS

Email: info@cesmii.org

CLEAN ENERGY SMART MANUFACTURING INNOVATION INSTITUTE™

https://twitter.com/SMCoalition

Facebook.com/SMLeadershipCoalition

http://www.linkedin.com /groups /Smart-Manufacturing-Leadership-Coalition

509493 Real-Time Adaptive Control of Carbon Fiber Production

Abstract:

Efficient integration of advanced sensors, data management, modeling and controls presents both tremendous challenges and opportunities in industry today. While these elements are all present in todayâ€™s manufacturing environment, they are largely implemented serially and independently resulting in inefficiencies when modifications are introduced in manufacturing systems or when engineering activities are performed to adapt manufacturing lines to changing conditions. This presentation describes ongoing efforts at Oak Ridge National Laboratory to develop a â€˜Smart Manufacturingâ€™ framework that couples technologies, software systems, and engineering workflow to optimize the manufacturing and enhance the understanding of Carbon Fiber production.

506183 Smart and Advanced Manufacturing Innovation in DOE

Abstract:

Smart manufacturing is an emerging field that combines physical systems with big data and intelligence. It is also referred to as Advanced Manufacturing, Digital Manufacturing, Cyber Physical Systems (CPS), the Internet of Things (IoT) and in Germany "Industrie 4.0." It can mean different things to different people, but essentially it is about smartly extracting information from the system to improve overall efficiency of manufacturing. Smart Manufacturing provides effective and secure human-system platform for better decision making and improving the overall productivity and efficiency of manufacturing across the networked enterprise. The talk will describe the objective of the "smart" manufacturing and the Clean Energy Smart Manufacturing Innovation Institute (CESMII). The talk will cover, among other things, the critical aspects of smart manufacturing that includes, reference architecture, verification, validation and uncertainty quantification, standards and protocols, and workforce training and education.

506185 The Technology and Vision of Smart Manufacturing

Abstract:

CESMIIâ€™s vision is that SM becomes so pervasive that by 2030, it is the next-generation way of doing manufacturing. Currently, both process-based and fabrication-based manufacturing are constrained by market and operational compartmentalization, slow change, vendor lock-in, and innovation by and for a few. By developing and accelerating broad-based innovations that are accessible by all, CESMII can be the driving force for moving the industry and market into a new, sustainable business model. In the broadest sense, CESMIIâ€™s end game is not about just cutting manufacturing costs or reducing energy consumption. Rather, the Institute is focused on enhancing U.S. manufacturing productivity by making better use of energy, materials, facilities, and workforce resources, which will improve the global competitiveness and reinvestment of U.S. manufacturing.

499589 A General Adsorption/Reaction Framework for Modular and Multi-Functional Process Design

Abstract:

Modular chemical processes are easily deployable for small scale operations in remote locations to produce value-added chemicals from unconventional resources and distributed feedstocks such as stranded natural gas, CO_2 from flue gas and associated shale gas. The advantages of modular processes include compactness, prevention of inefficiencies of on-site equipment manufacturing, freedom of constructing high-quality equipment in specialized locations, and faster production time. However, modular processes often fail to achieve economies of scale. One way to address this challenge is to design *multifunctional* modular processes using existing single-functional materials. To this end, we first propose a multi-material design approach which integrates multiple phenomena within a single unit and thereby leads to multifunctional modular processes using existing materials. We use a systematic framework to select and arrange materials (e.g., adsorbent and catalyst blocks) with unique tasks (e.g., adsorption, reaction, storage) in a sequential block-by-block arrangement inside a packed column, and configure operational cycles for integrated, periodic and multifunctional processes. The task-integrated column operates in a cyclic or periodic manner, where each cycle consists of several steps. We use a single superstructure that embeds different design alternatives using blocks, where each block represents a specific material performing a specific task for a specific time step. The functional block-based superstructure is general such that it contains numerous cyclic process configurations for separation processes including pressure/temperature swing adsorption (PSA/TSA), as well as for integrated separation-conversion processes including sorption-enhanced reaction processes (SERP).

Next, in order to accurately predict the performance of different processes generated via multi-material design, we develop Generalized Reaction-Adsorption Model (GRAM) â€“ a high-fidelity process model for dynamic adsorption-reaction systems. GRAM is a one-dimensional, pseudo-homogeneous, non-isothermal, non-adiabatic, and non-isobaric model that captures the reaction and adsorption dynamics in hybrid adsorption-reaction systems. The model is validated using different experimental data from literature. Specifically, the model predictions show agreement with experimental observations for sorption-enhanced steam-methane reforming (SMR)[1,3], sorption-enhanced reverse water gas shift (RWGS)[2], and conventional SMR[4,5]. Computationally, the model handles multi-feed, multi-functional and multi-product processes well.

The GRAM model is further incorporated within the multi-material design framework to design several novel multifunctional modular processes for cost-effective carbon capture and utilization using unconventional energy sources such as stranded and distributed natural gas, CO_2-contaminated shale gas, biogas, landfill gas, fuel gas and flare gas. The objective is to produce valuable products such as hydrogen and syngas. Computational studies are performed for several configurations of the column containing (i) pure catalyst, (ii) pure adsorbent, (iii) heterogeneously-compartmentalized adsorbent and catalyst, and (iv) homogeneously-distributed uniform mixture of adsorbent and catalyst. These intensified processes use sections of adsorbent and catalyst for simultaneous capture and conversion of reactants.

References:

[1] Carvill, B.T., Hufton, J.R., Anand, M., Sircar, S. Sorption-Enhanced Reaction Process. AIChE Journal. 1996.

[2] Jang, H.M., Lee, K.B., Caram, H.S., Sircar, S. High-purity hydrogen production through sorption enhanced water gas shift reaction using K_2CO_3-promoted hydrotalcite. Chemical Engineering Science. 2012.

[3] Hufton, J.R., Mayorga, S., Sircar, S. Sorption-enhanced reaction process for hydrogen production. AIChE Journal. 1999.

[4] Shu, J., Grandjean, B.P.A., Kaliaguine, S. Methane steam reforming in asymmetric Pd- and Pd-Ag/porous SS membrane reactors. Applied Catalysis. 1994.

[5] Gallucci, F., Paturzo, L., Basile, A. A simulation study of the steam reforming of methane in a dense tubular membrane reactor. International Journal of Hydrogen Energy. 2004.

496493 Ultrasonic Tuning of TiO_2 Based Mixed Oxides Structural Properties and Catalytic Activity

Abstract:

Ultrasound (US) has a potential to impact every stage of the preparation of a material in liquid phase. US can produce nanomaterials in amorphous state (due to the very high cooling rates), which is relevant to fields such as catalysis, magnetism and coating processes [1]. US generates shock waves that facilitate coating and insertion/intercalation processes and improves the distribution of the active phase on a support. We pursue ultrasound powder synthesis as a greener method to lower solvent amount and catalyst precursor, as well as to accelerate lengthy preparation method such as sol-gel and impregnation. US-assisted synthesis produces catalysts with better morphology and activity. Simple tuning of operating conditions increases the range of obtainable features for different purposes [2,3].

We report the ultrasound assisted sol-gel and impregnation synthesis of sulphated ZrO_2-TiO_2 and Mn oxides-TiO_2 systems. We characterized the catalysts with N_2 adsoption, XRD, SEM-EDX and TEM. We tested the sulphated ZrO_2-TiO_2 in the free fatty acids (FFA) esterification reaction and the sulphated ZrO_2-TiO_2 and the Mn_xO_y-TiO_2 systems in the ultrasound-assisted photocatalytic degradation of recalcitrant water pollutants (pharmaceutical products). We adopted a ultrasound probe with a tip of 13 or 20 mm diameter. The ultrasound operating frequency was fixed at 20 KHz. The nominal power was 500 W and we varied the amplitude from 20 to 60 % adopting continuous ultrasound or ultrasound pulses (0.5 on and 0.5 off). In the sol-gel synthesis of sulphated ZrO_2-TiO_2 systems, ultrasound pulses increased the surface area up to 220 m^2g^{-1} for the lowest solvent: precursors ratio of 15:1 mol, corresponding to the highest power density. Solvent: precursors molar ratios of 30:1 and 60:1 gave a SSA of 150 m^2g^{-1} and 30 m^2g^{-1}, respectively. The samples obtained with pulses were also the ones exhibiting the highest Brønsted surface acidity. The catalysts with the highest acidity/SSA ratio were the most active in the free fatty acids esterification reaction.

We deposit Mn_xO_y on TiO_2 in one step with the ultrasound-assisted sol-gel synthesis or via ultrasonic impregnation in a second step. We prepared Mn/TiO_2 catalyst in one step with a classical sol–gel method, starting from butyl-titanate, deionized water, nitric acid, anhydrous ethanol, and manganese nitrate [4]. After mixing at room temperature to yield a yellowish transparent sol, we dried the precursor at 80°C for 24h, converting it into a xerogel. The calcination to obtain the Mn/TiO_2 catalyst was standard at 500°C for 5h. We prepared the same catalyst applying US, either maintaining the same calcination parameters or decreasing the calcination temperature to 400°C for 2 hours to trigger the formation of anatase instead of rutile as an active phase.

Mn/TiO_2 two steps synthesis consisted of the US-assisted surface deposition of Mn-oxides on TiO_2 (both nanometric or micrometric), synthesized in a preceding stage [5]. Generally, US promote homogeneous nucleation in a first step, and enhance the crystals growth in a second stage. Acoustic cavitation, in particular for the two-steps Mn deposition, generates high-speed micro jets which impact the TiO_2 support with a cleaning and de-agglomerating effect. Ultrasound increases the dispersion of the Mn species over both supports and has a role in stabilizing Mn_xO_y species that samples synthesized with traditional sol-gel don't normally exhibit (preliminary XRD results) [6]. The modulation of the main sonication operating parameters produces a mixed-size distribution of Mn and Mn-Ox species on TiO_2 rather than a narrow distribution of small crystallites, which often turns out to be better for the photocatalytic activity. US also improves the distribution of Mn_xO_y over micro-metric TiO_2 (SEM-EDX analysis), which is meaningful from a safety point of view since body barriers may not "recognize" nano-powders. The photocatalytic activity of Mn_xO_y over nano and micro-size TiO_2 is comparable. Considering instead the comparison between Mn and other metals, Mn is cheaper and one of a kind, considering the variety of oxides that can form. By means of US we can also modulate and enhance the formation of specific oxidation states [6,7].

References

[1] T.J. Mason, Ultrasound in synthetic organic chemistry, *Chem. Soc. Rev.*, 26 (1997), pp. 443–451.

[2] S.J. Musevi, A. Aslani, H. Motahari, H. Salimi, Offer a novel method for size appraise of NiO Nanoparticles by PL analysis: synthesis by sonochemical method, *J. Saudi Chem. Soc.*, 20 (2016), pp. 245–252.

[3] C. Pholnak, C. Sirisathitkul, D.J. Harding, Characterizations of octahedral zinc oxide synthesized by sonochemical method, *J. Phys. Chem. Solids*, 72 (2011), pp. 817–823.

[4] Qi-lin Chen, Rui-tang Guo, Qing-shan Wang, Wei-guo Pan, Ning-zhi Yang, Chen-zi Lu, Shu-xian Wang, *J. Taiwan Inst. Chem. Eng.*, 64 (2016) 116–123.

[5] M. Stucchi, C.L. Bianchi, C. Pirola, G. Cerrato, S. Morandi, C. Argirusis, G. Sourkouni, A. Naldoni, V. Capucci, *Ultrason. Sonochem.*, 31 (2016) 295–301.

[6] M.I. Díez-García, V. Manzi-Orezzoli, M. Jankulovska, S. Anandan, P. Bonete, R. Gómez, T. Lana-Villarreal, Effects of ultrasound irradiation on the synthesis of metal oxide nanostructures, *Phys. Procedia*, 63 (2015), pp. 85–90.

[7] K.D. Bhatte, D.N. Sawant, D.V. Pinjari, A.B. Pandit, B.M. Bhanage, One pot green synthesis of nano sized zinc oxide by sonochemical method, *Mater. Lett.*, 77 (2012), pp. 93–95.

497703 Multi-Model Operability Approach for Process Design, Intensification and Modularity: Application to Nonlinear and High-Dimensional Membrane Reactors

Abstract:

Membrane reactors (MRs) have arisen as alternatives for conventional reactors, enabling previously economically unfavorable processes. MRs are also intensified processes that are candidates for modular applications. Such reactors provide a built-in membrane separation layer along the entire reactor length, where one or more reaction products can be selectively removed. This removal shifts the reaction equilibrium towards the products, increasing reactants conversion and production rate. However, the process design of MRs is intrinsically challenging. MRs combine chemical reactions with separation transport phenomena, which results in models of complex and nonlinear nature for their representation. MRs are also commonly placed in highly constrained environments, which poses an additional difficulty to the process design task, especially if process integration is needed for modularity. In this presentation, a multi-model operability approach is proposed to address these challenges. Operability was first introduced as a tool for the design and control interface to calculate the inputs that are needed to achieve specified desired outputs in a process [1]. Here, the operability pathway is extended to provide a framework able to tackle nonlinear high-dimensional and constrained systems. This framework attains process intensification, ensuring high efficiency units, suitable for modular systems applications.

In the past decades, process operability has been traditionally applied to low-dimensional nonlinear systems. The implementation of operability for high-dimensional nonlinear systems was successfully performed by employing optimization algorithms based on nonlinear programming (NLP) [2,3]. The novelty of this work derives from the fully utilization of linear programming (LP) tools to handle systems of the same complexity, with the advantage of computational time reduction. In particular, the original nonlinear system is divided into several subsystems and a linear model is calculated for each subsystem. The properties of all linear models are then analyzed, such as the eigenvalues (or singular values), so that the minimum number of model sets that adequately represents the process nonlinearity is obtained. A multi-level optimization algorithm is then formulated based on LP concepts, aiming intensification and control targets, while considering process requirements and constraints.

The developed framework is applied to a membrane reactor for the direct methane aromatization conversion (DMA-MR) to hydrogen and benzene. DMA-MR corresponds to a candidate system for modular natural gas utilization onsite at stranded gas fields without the need of building expensive pipelines [2]. Preliminary results for a low-dimensional system show that benzene production can be optimized and the membrane reactor footprint reduced by 60% in terms of volume (catalyst zone) and 80% in membrane area. These results indicate the potential of the proposed framework to facilitate process intensification towards modularity. In this presentation, a comparison will be established between this approach and previously developed NLP-based approaches [4] to analyze the trade-off between accuracy and computational expense. The expansion to a higher dimensional system will then be discussed, contemplating membrane parameters, design and operating conditions that are critical for process intensification and modularity.

References:

1. Vinson D. R. and Georgakis C. "A new measure of process output controllability". J. Proc. Cont. 2000, 10(2-3), 185-194.
2. Carrasco J. C. and Lima F.V. "An optimization-based operability framework for process design and intensification of modular natural gas utilization systems". Accepted for publication in Comput. Chem. Eng. 2017; DOI: CACE-5645.
3. Carrasco J. C. and Lima F. V. "Novel operability-based approach for process design and intensification: application to a membrane reactor for direct methane aromatization". AIChE J. 2017; 63(3): 975-983.
4. Carrasco J.C. and Lima F. V. "Operability-based approach for process design, intensification, and control: application to high-dimensional and nonlinear membrane reactors". In proceedings of the FOCAPO/CPC, 2017.

493441 A Modeling Methodology for Predicting Intensified CO_2 removal Efficiency with Rotating Packed Bed

Abstract:

The opportunity in process intensification (PI) of carbon dioxide (CO_2) capture has drawn much attention recently. Since last decade, rotating packed bed (RPB) has been considered as an intensified alternative to conventional packed bed (PB) in CO_2 chemical absorption processes which require excessive absorber volume resulting from mass transfer resistance across vapor-liquid interface during the absorption process [1-3].

Reliable predictions of intensified CO_2 removal efficiency in RPB are essential to design the intensified CO_2 absorption/stripping process. In this study, we present a modeling methodology with sound theoretical basis for predicting the CO_2 removal efficiency in RPB. The methodology assumes the modeling of RPB and that of PB are identical if the gravitational force presented in RPB can be properly taken into account in modeling the mass transfer resistance. The methodology consists of three steps. First, we validate the PB model with PB experimental data [4]. The PB model is based on Aspen rate-based PB model with liquid film discretization (Aspen RateSep) [5]. Second, we predict the CO_2 removal efficiency of RPB experiments [1,4] with the validated PB model. Through this step, we identify the difference in CO_2 removal efficiency with and without the gravitational effect. Third, we account for the gravitational effect in RPB with the adjustable mass transfer factors in the validated PB model. Specifically, the liquid mass transfer coefficient factor and the interfacial area factor are chosen to account for the RPB gravitational effects with the validated PB model. The absolute average deviation (AAD) between the prediction results and the experimental RPB data [1,4] are found to be less than 10% when lean loading of MEA is in the typical operating range of 0.16 ~ 0.44 mol CO_2/mol amine, suggesting a robust predictive capability of our modeling methodology.

This modeling methodology not only provides new insights on RPB, but also avails chemical engineers a practical and theoretically sound modeling capability to estimate the intensified CO_2removal efficiency when replacing PB with RPB. Furthermore, by taking advantages of proven modeling tools in commercial simulators, the methodology could be widely practiced by engineers worldwide to examine the process intensification potential of replacing PB with RPB.

[1] Yu, C.H., Chen, M.T., Chen, H., Tan, C.S., "Effects of Process Configurations for Combination of Rotating Packed Bed and Packed Bed on CO_2 Capture", *Applied Energy*, 175, 269-276 (2016).

[2] Yu, C.H., Lin, Y.X., Tan, C.S., "Effects of Inorganic Salts on Absorption of CO_2 and O_2 for Absorbents Containing Diethylenetriamine and Piperazine", *Int. J. Greenh. Gas Cont.*, 30, 118-124 (2014).

[3] Yu, C.H., Cheng, H.H., Tan, C.S., "CO_2 Capture by Alkanolamine Solutions Containing Diethylenetriamine and Piperazine in a Rotating Packed Bed", *Int. J. Greenh. Gas Cont.*, 9, 136-147 (2012).

[4] Thiels, M., Wong, D.S.H., Yu, C.H., Kang, J.L., Jang, S.S., Tan, C.S. Modeling and Design of Carbon Dioxide Absorption in Rotating Packed Bed. *11th IFAC Symposium on Dynamics and Control of Process Systems.*895-900 (2016).

[5] Zhang, Y., Chen, H., Chen, C.C., Plaza, J.M., Dugas, R., Rochelle, G.T. Rate-Based Process Modeling Study of CO_2 Capture with Aqueous Monoethanolamine Solution. *Ind. Eng. Chem. Res.*48, 9233-9246 (2009).

Keywords: Process Intensification; Rotating Packed Bed; Aspen Plus; CO_2 Chemical Absorption; Mass Transfer Coefficient; Interfacial Area

482122 Carbon Dioxide Capture By Sodium Hydroxide-Glycerol Aqueous Solution in a Rotating Packed Bed

Abstract:

Carbon dioxide capture by chemical absorption has been one of the most popular approaches. However, low solubility of CO_2 in water and high energy demand for absorbent regeneration hinder its feasibility. In this study, glycerol, a biodiesel waste by-product, was added to sodium hydroxide (NaOH) aqueous solution to relax these drawbacks, and to serve as an efficient absorbent in a rotating packed bed (RPB) for CO_2 capture. The experimental results suggested that the addition of glycerol in the NaOH aqueous solution boosted the overall CO_2 absorption efficiency to more than 90% easily. Under the high G (rotating speed) condition, the (chemical) enhancement factor could go up to more than 50. Furthermore, at same NaOH concentration, the mass transfer coefficients (K_Ga) could be doubled even when the viscous glycerol was added because of intensified centrifugal field. It is clear indicated that, in an RPB, glycerol addition in NaOH solution showed very positive effect on CO_2 capture not only for better mass transfer performance but also for the potential utilization of biodiesel by-product.

484778 Graphic Synthesis Method for Multi-Technique Integration Separation Sequences of Multitudinous Refinery Gases

Abstract:

Graphic Synthesis Method for Multi-technique Integration Separation Sequences of Multitudinous Refinery Gases

Xuehua Ruan, Hongyan Xiao, Xiaobin Jiang, Xiaoming Yan, Yan Dai, Gaohong He

School of Petroleum and Chemical Engineering, State Key Laboratory of Fine Chemicals, Dalian University of Technology, Panjin, Liaoning, P.R. China, 124221

Corresponding Author Email: hgaohong@dlut.edu.cn

In petroleum and chemical industries, considerable crude oil, i.e., 3~8 %, is converted into refinery gases and used as furnace fuels. Many valuable species in the refinery gases, e.g., H_2, C_2H_6, C_2H_4, C_3H_8 and C_4H_{10}, are depleted. Currently, this wastage is going on increasing because more and more hydro-processes are built up for refining heavy sour crude oils to high quality products. Accordingly, the effective separation of valuable species from refinery gases has been one of the most potential approach to improve crude oil utilization percentage.

Nowadays, some separation unit operations have been attempted to fractionate and reclaim refinery gases, e.g., pressure swing adsorption, glassy polymer membrane permeation and cryogenic condensation for recovering H_2, shallow condensation, gasoline absorption and rubbery polymer membrane permeation for concentrating light hydrocarbons, and distillation for further separating hydrocarbons. Nevertheless, any simplex technique cannot fractionate all the valuable components efficiently and sufficiently from multitudinous refinery gases. It is necessary to integrate various techniques together as the hybrid processes to achieve the recovery task. In this instance, a shortcut method is established in this work to construct the multi-technique integration processes, in which highly efficient separation unit operations are coupled together within an optimal sequence to improve efficiency.

This shortcut synthesis method is a graphic means based on expressing separation units by vectors in composition space. Three supporting technical measures are summarized:

1. After ranking components with their separation characteristics and belongings in products, refinery gases are reduced to be pseudo-ternary mixtures of H_2, light hydrocarbons and fuel gases, meanwhile the composition space is visualized into triangular coordinate system.

2. Based on the efficiency analysis, dominant feed ranges for adsorption, condensation, glassy and rubbery membrane separation are determined, in which the represented technique could be more efficient than the others for the composition matched refinery gases.

3. Separation operations, usually acting with the selectivity unique to a certain species group, are estimated quickly by the shortcut calculations and then represented with vectors in the triangular coordinate system.

The shortcut synthesis method and its application for construct multi-technique integration systems are sketched out in Figure 1.

Figure 1. Graphic synthesis of a multi-input cascade coupling separation sequence.

The synthesis is launched by positioning the feedstock in the triangular coordinate system according to the composition. According to the location of the feed stream, the separation unit suitable to deal with this stream can be determined facilely by matching with the dominant ranges. Afterwards, the shortcut calculations are conducted by vector analysis to orientate the residue from a separation unit and select the next separation unit.

Thanks to running the separation unit operations in their preferred ranges and avoiding the stream mixing with great composition difference, the multi-technique systems designed by the expert synthesis approach can reclaim refinery gases sufficiently and efficiently. For multitudinous refinery gases in large-scale petrochemical complex, the overall recovery ratio for valuable species is higher than 95 %, while the specific energy consumption is usually lower than 0.50 kWh/Nm^3 refinery gases. The integration processes designed by this means have been successfully industrialized in Sinopec and PetroChina.

Acknowledgement:

The authors are grateful to the financial support from National Natural Science Foundation of China (No. 21606035, U1663223, 21527812), Education Department of the Liaoning Province of China (LT2015007), and Changjiang Scholars Program (T2012049).

500110 Innovative Design and Systematic Process Intensification Using Building Blocks

Abstract:

Process intensification addresses the development of techniques that leads to substantially smaller, cleaner, safer, and more energy-efficient technologies. It also enables the innovative design of novel multi-functional units (e.g., Eastman Chemical methyl acetate process that performs extractive distillation, reactive distillation, reaction and distillation within a single column [1,2]). Intensification methodologies have been developed in the past, several of which are described in [3-8]. In this work, we address the following design and intensification challenges for both steady-state and dynamic processes: (i) how to systematically select and combine multiple phenomena to create new opportunities and obtain "out-of-the-box" design solutions, (ii) how to systematically identify intensified equipment and, at the same time, generate process flowsheets containing these equipment, (iii) how to select optimal configurations of intensified equipment and process flowsheets, and (iv) how to incorporate process intensification into process synthesis and optimization without *a priori* postulation of intensification alternatives. To address these questions within a single framework, we depart from the classical unit operation-based representation of process units, flowsheets and superstructures, and propose a new representation using finite building blocks [9]. In this representation, a physiochemical phenomenon, a processing task or a function can be represented using a single block or using two neighboring blocks and their boundaries. Other representations also exist in literature, including the modular [10,11] and phenomena building blocks (PBBs) [6,7]. Here, we propose a novel and general process superstructure which is an ensemble of only building blocks in two dimensions. Each block is denoted by its coordinate within the superstructure. A block can have entering and outgoing streams in both horizontal and vertical directions. An assembly of blocks of the same type obtains a classical unit, while an assembly of blocks with different types results in an intensified unit. The block superstructure encompasses numerous process equipment and flowsheet alternatives, and can be used to design, synthesize and intensify a chemical process without *a priori* postulation of these alternatives. The overall problem is formulated using a single mixed-integer nonlinear optimization (MINLP) model.

For dynamic intensification of multi-functional and periodic/cyclic processes, we propose a similar block superstructure, where each block represents a specific material performing a specific task for a specific time step. The superstructure is specifically suitable for the design of integrated, multi-functional and modular adsorption-reaction systems embedded within a single column while exploiting the dynamic interactions and trade-offs between different phenomena. Again, from the same superstructure, one can systematically generate variants of adsorption-reaction systems, including single- or multi-layered packed adsorption columns, reactors, pressure/temperature swing adsorption (PSA/TSA) processes, pressure and temperature swing adsorption (PTSA) processes, sorption-enhanced reaction processes (SERP), simulated moving bed chromatography (SMB) processes, and reactive SMB processes. Using this method, various intensified and modular processes can be designed for the utilization of both conventional and unconventional resources including stranded natural gas, contaminated shale gas, fuel gas or landfill gas to produce sustainable fuels and chemicals [12].

References:

[1] Agreda, V.H., Partin, L.R. (1984). Reactive distillation process for the production of methyl acetate. U.S. Patent 4,435,595.

[2] Siirola, J. J., (1996). Industrial applications of chemical process synthesis. Advances in Chemical Engineering 23, 1–62.

[3] Stankiewicz, A. I., Moulijn, J. A. (2000). Process Intensification: Transforming Chemical Engineering. Chemical Engineering Progress 1, 22–34.

[4] Lutze, P., Gani, R., Woodley, J. M. (2010). Process intensification: a perspective on process synthesis. Chemical Engineering Processing: Process Intensification 49 (6), 547–558.

[5] Reay, D., Ramshaw, C., Harvey, A. (2013). Process Intensification: Engineering for Efficiency, Sustainability and Flexibility. Butterworth-Heinemann, Waltham, MA.

[6] Babi, D. K., Cruz, M. S., Gani, R. (2016). Fundamentals of process intensification: a process systems engineering view. In: Segovia-Hernández, J.G., Bonilla-Petriciolet, A. (Eds.), Process Intensification in Chemical Engineering., http://dx.doi.org/10.1007/978-3-319-28392-0 2.

[7] Tula, A. K., Babi, D. K., Bottlaender, J., Eden, M. R., Gani, R. (2017). A computer-aided software-tool for sustainable process synthesis-intensification. *Computers & Chemical Engineering*. http://doi.org/10.1016/j.compchemeng.2017.01.001.

[8] Baldea, M. (2015). From process integration to process intensification. *Computers & Chemical Engineering, 81*, 104-114.

[9] Demirel, S. E., Li, J., and Hasan, M. M. F., 2017. Systematic Process Intensification using Building Blocks, *Computers and Chemical Engineering*, http://dx.doi.org/10.1016/j.compchem eng.2017.01.04

[10] Papalexandri, K. P., Pistikopoulos, E. N. (1996). Generalized modular representation framework for process synthesis. *AIChE Journal, 42*(4), 1010-1032.

[11] Ismail, S. R., Proios, P., Pistikopoulos, E. N. (2001). Modular synthesis framework for combined separation/reaction systems. *AIChE Journal, 47*(3), 629-649.

[12] Iyer, S. S., Bajaj, I., Balasubramanian, P., Hasan, M. M. F. Modular Process Intensification of Carbon Capture and Conversion to Syngas. Submitted.

501702 Modeling and Simulation of a Hybrid Adsorptive-Membrane Reactor (HAMR) for Intensification of the Water-Gas Shift (WGS) Reaction Process

Abstract:

The Hybrid Adsorptive-Membrane Reactor (HAMR) is a dynamically operated process, which consists of a metal tube containing one or more tubular membranes. The HAMR's reaction section, surrounding the membrane tubes, contains catalyst and adsorbent material that accelerate reaction kinetics and adsorb some reaction products from the reacting mixture respectively. The HAMR's permeation section consists of the membrane tubes' interior, and physically removes some other reaction products from the reacting mixture. The HAMR is typically operated in reaction and regeneration modes, since the adsorbent meterial reaches its adsorption capacity after a period of operation, and thus needs to be regenerated.

In this work a mathematical model of HAMR is developed and simulated. The aim of this effort is to intensify the water gas shift (WGS) reaction process for hydrogen production. Indeed, the coupling of reaction and separation in the HAMR process can substantially improve reactant conversion, product selectivity, and can even yield near pure products. In addition, for reversible reactions such as the WGS, the HAMR process can be designed so as to deliver reactant conversions that exceed even equilibrium conversions attainable in conventional reactors. The HAMR system is composed of a, where some reaction products permeate. The proposed HAMR model quantifies velocity, species concentration, and temperature profiles throughout the reaction-adsorption, and permeation zones, by solving momentum/species/energy transport equations accounting for convection/reaction /diffusion/conduction mechanisms.

The developed model is used to intensify the Water Gas Shift Reactor (WGSR) Process. To this end, the performance of the traditional packed bed reactor is first quantified. Then parametric studies of the HAMR are carried out, so as to identify maximum intensification designs. These studies include a broad range of operating conditions and parameters (e.g. reactor operating temperature, catalyst and adsorbent weight to feed flowrate ratios, and others.

496925 Intensification of Chemical Processes Via Periodic Operation

Abstract:

Process intensification has emerged over the past few decades as a framework for increasing process performance by bringing chemical and physical phenomena into closer physical proximity. Intensification often transcends the traditional unit operations framework, with several conventional operations occurring in the same physical device. While this minimizes transport limitations and reduces equipment size, it can also lead to control challenges: intensified processes feature strong dynamic coupling between phenomena and have fewer control degrees of freedom than their conventional counterparts[1].

In this presentation, we describe a new means of operating intensified processes, focusing on deliberately imposing a transient (specifically, periodic) operating regime in order to improve average process performance. Additionally, we discuss periodic operation in terms of expanding the number of control degrees of freedom, in the sense that the amplitude and period of oscillation of a single manipulated variable can be set independently.

Using a prototype process featuring several coupled (and possibly competing) phenomena, we demonstrate theoretically that periodic operation can increase overall economic performance. We perform simulations to illustrate our theoretical findings, using both the canonical example described above, and the model of a dividing-wall column.

References

[1] Schembecker, G.; Tlatlik, S. Process Synthesis For Reactive Separations. Chem. Eng. Process. 2003, 42, 179-189.

498144 Intensification of the Hydride Vapor Phase Epitaxy Manufacturing Process for Solar Devices

Abstract:

Photovoltaics made of III-V compound semiconductors, especially in the form of multi-junctions, have demonstrated great conversion efficiencies [1]. However, their application is limited significantly by the expensive fabrication methods and materials. To address this challenge, hydride vapor phase epitaxy (HVPE) has regained research interest because of its intrinsically higher growth rate and lower precursor costs compared with other epitaxial growth methods. But the high growth rate and near-thermodynamic-equilibrium nature of this process also impose difficulties of producing large-area uniform wafer in a controlled way. The process needs to be geared into a more economic and manageable configuration to produce high quality products with stable properties at rapid throughput.

There are already a couple of practices applying the philosophy of process intensifications in this area. For example, through clever design, the compact horizontal reactor in [2] was capable of growing various III-V semiconductors (GaAs, InGaAs, GaInP, etc.) at high growth rates and depositing multilayers of those compounds in a controlled mode by manually operating transfer rod and shifting gas flow [2, 3]. Based on this reactor, an improved inline reactor was proposed [4], in which the growth method was adapted to sequentially deposit different layer of material on one wafer travelling through multiple reaction chambers separated by gas curtain. To further promote performance of process and product, we present novel alternatives to previous intensification techniques using new concepts in both equipment design and operation method. Computational modeling tools are used to verify design and operational strategies. We are expecting this work will not only aid the production of economically and industrially viable III-V solar cells, but also inspire more ideas to optimize other CVD processes that grow films with high cost-performance ratio.

[1] M. A. Green et al., "Solar cell efficiency tables (version 49)," Prog. Photovolt: Res. Appl., vol. 25, no. 1, pp. 3–13, Jan. 2017.
[2] K. L. Schulte, W. L. Rance, R. C. Reedy, A. J. Ptak, D. L. Young, and T. F. Kuech, "Controlled formation of GaAs pn junctions during hydride vapor phase epitaxy of GaAs," Journal of Crystal Growth, vol. 352, no. 1, pp. 253–257, Aug. 2012.
[3] K. L. Schulte et al., "Metalorganic vapor phase growth of quantum well structures on thick metamorphic buffer layers grown by hydride vapor phase epitaxy," Journal of Crystal Growth, vol. 370, pp. 293–298, May 2013.
[4] D. L. Young, A. J. Ptak, T. F. Kuech, K. Schulte, and J. D. Simon, "High throughput semiconductor deposition system," US20130309848 A1, 21-Nov-2013.

501295 Process Intensification of Reactive Separator Networks through the Ideas Conceptual Framework

Abstract:

A method to rigorously identify the performance limits of a reactive separator network is presented in this work. The quantification of the enhancement potential for a given technology can greatly benefit process intensification studies in the pursuit of radical improvements. The Infinite DimEnsionAl State-Space (IDEAS) conceptual framework is first reviewed and then shown to be capable of assessing the potential for process intensification of reactive separation processes. The IDEAS framework is employed to formulate an infinite linear program (ILP) that can synthesize optimal reactive separation networks, and establish rigorous tradeoffs between total network reactive holdup, and total network capacity. The proposed process intensification method is demonstrated on a case study involving the metathesis of 2-pentene through reactive distillation, an important chemical process for the oil industry. Total reactive holdup and total capacity, a surrogate for the amount of catalyst used and capital cost respectively, were selected as the systemâ€™s variables of interest for process intensification. Using a traditional reactor-followed-by-separation-system scheme as a baseline, an intensified IDEAS based reactive separation network can reduce the total reactive holdup and total capacity by 94.5% and 84.3% respectively. Results available in the literature were converted to the total holdup-total capacity space in the extent possible and compared with the IDEAS results. The comparison shows the existence of improvement potential, indicating that process intensification opportunities can be further explored in this reactive separation system.

490321 Intensified Reaction-Separation Schemes

Abstract:

Intensified reaction-separation schemes

Evangelia Koumaditi, John Woodley, Rafiqul Gani *Department of Chemical and Biochemical Engineering, Technical University of Denmark, DK-2800 Lyngby, Denmark*

The environmental and economic concerns are driving the chemical and biochemical industries to develop new, sustainable methods of production. Combining the reaction and the separation process in one unit, not only can it lead to cost and energy reductions for the overall process, but it can also have an effect on the reaction, in terms of yield. Motivated by the need to take advantage of the intensified process units, a method needs to be developed, which will allow the systematic generation and analysis of reactive separation configurations at an early stage in process synthesis. The objective of this work is the development of a method to design reaction-separation schemes, based on analysis of phase equilibrium of the involved chemical system.

Oxidation reactions are used as a generic system to highlight the conceptual basis of the design method. The selective oxidation of primary alcohols are examples of this generic system, which are used as case studies for the application of the design method. The properties of the pure components and the characteristics of phase equilibrium (azeotropes and liquid phase split) have been considered to identify new and novel intensified reaction-separation schemes. Also, the selective oxidation of alcohols to aldehydes is a challenging reaction, as aldehydes are relatively unstable products and also due to their use in the production of high-value products. Even though the reactions are considered to be bio-catalyzed by the enzyme Galactose Oxidase, the design method nevertheless is applicable to other similar systems and also not restricted to the use of biocatalysts only.

The specific examples highlighted in this work cover ternary systems involving alcohol + water + aldehyde, from which aldehyde is the product to be recovered, while alcohol and water are recycled back to the reactor. The first step of the method is the definition of the reaction system and data collection. The designs of the intensified separation schemes are based on the properties of pure components, as well as the properties of the mixture of the involved chemical system. The properties of pure components that are of interest are namely, the boiling point, the vapor pressure and the water solubility. After the pure component analysis is completed, the vapor-liquid and liquid-liquid equilibria are examined for each binary subsystem of the ternary mixture, employing available data or generated through model-based phase equilibrium predictions. The next step involves the classification of the systems according to their characteristics. The reaction-separation schemes are then determined by using phase equilibrium relations that favor the removal of the product from the reacting phase. Calculations of single-stage equilibrium based separations are performed in order to obtain the mass balance and the distribution of the compounds in the different phases. The last step involves validation tests to confirm the feasibility of the suggested configuration, by super-imposing the calculated mass balance lines on the predicted VLE and LLE plots. Success of the designed intensified reaction-separation scheme depends on the amount of the targeted compound removed from the reactor. Note that the separation and recovery of the product requires the use of solvents. Ionic liquids, because of their special properties have been considered as the needed solvents.

The presentation will highlight the results of the intensified reaction-separation schemes for a class of oxidation of primary alcohol reactions. Details of the design method together with the important concepts will also be highlighted.

499187 Process Intensification of Large-Scale Continuous Biobutanol Production via a Multi-Feed Bioreactor with *in situ* Gas Stripping

Abstract:

The economical production of alternative liquid fuels is vital to achieving energy independence from fossil fuel sources. While ethanol is often considered the primary candidate, butanol offers the advantages of higher energy density, lower vapor pressure, and lower volatility [1]. However, butanol is more toxic to cell cultures than ethanol, making it difficult to produce biologically at levels necessary for economic viability. Process intensification has been used in bioreactors by combining acetone, butanol, and ethanol (ABE) fermentation with separation phenomena, including pervaporation and gas stripping, to selectively remove the ABE products and overcome this limitation [2, 3]. Additionally, multi-feed inlet configurations have been investigated in continuous bioprocessing systems to limit inhibitory effects in bioreactors by manipulating the substrate concentration and dilution rate independently, limiting substrate concentration in the reactor without necessarily decreasing the reactor productivity [4]. The development of an intensified bioreactor with a multi-feed inlet configuration has the potential to overcome the toxicity limitation, increase process throughput, and produce butanol in a continuous, economically attractive manner that can compete with current bioethanol processes.

In this work, we examine the economic viability of large-scale, continuous production of biobutanol using *Clostridium acetobutylicum* and *in situ* gas stripping. First, the optimal control problem for a multi-feed intensified bioreactor is investigated using the flowrate of the stripping gas and inlet feeds as manipulated variable to maximize profitability. Batch reaction kinetics for the system, taken from Votruba et al., are extended to continuous processing in the formulation of the optimal control problem [5]. Solutions for the intensified bioreactor are compared to the economic viability of an optimally-controlled, continuous biobutanol reactor in the absence of gas stripping to emphasize the necessity of intensified process operations.

Once the optimal control policy for the bioreactor is determined, the process synthesis problem for large-scale biobutanol production from lignocellulosic biomass is explored by maximizing profitability. The synthesis problem considers using multiple bioreactors in parallel to process glucose made available from five options of cellulosic biomass, hybrid poplar, corn stover, sorghum, sugarcane bagasse, and switchgrass, in combination with four possible pretreatment methods, ammonia fiber expansion (AFEX), dilute sulfuric acid, sodium hydroxide, and liquid hot water. The use of lignin as a source of steam and salable electricity generation is also considered. The results of this study look to emphasize the importance of process intensification practices in the areas of bioprocess to develop new and otherwise economically impractical processes.

1. Liu, X.B., et al., *Enhancement of butanol tolerance and butanol yield in Clostridium acetobutylicum mutant NT642 obtained by nitrogen ion beam implantation.* Journal of Microbiology, 2012. **50**(6): p. 1024-1028.
2. Qureshi, N. and H.P. Blaschek, *Recovery of butanol from fermentation broth by gas stripping.* Renewable Energy, 2001. **22**(4): p. 557-564.
3. Li, J., et al., *Efficient production of acetone-butanol-ethanol (ABE) from cassava by a fermentation-pervaporation coupled process.* Bioresource Technology, 2014. **169**: p. 251-257.
4. Raftery, J.P., M.R. DeSessa, and M.N. Karim, *Economic improvement of continuous pharmaceutical production via the optimal control of a multifeed bioreactor.* Biotechnol Prog, 2017.
5. Votruba, J., B. Volesky, and L. Yerushalmi, *Mathematical-Model of a Batch Acetone Butanol Fermentation.* Biotechnology and Bioengineering, 1986. **28**(2): p. 247-255.

503772 Process Systems Engineering Contributions in Pharmaceuticals

Abstract:

Healthcare is a major and rapidly growing sector of every developed economy, with 2015 world wide healthcare costs exceeding $7.7 trillion per annum. It is notable that US healthcare costs have risen to 18% of GDP. While the largest portion of the cost is for hospital, clinical, laboratory and physician services, some 10% is on expenditures for medicines. Although a modest portion of the total, the medicine component is receiving heightened attention due, first, to its immediate impact on patient pocketbooks but also to concerns about drug safety, quality and even availability. Our primary focus in this presentation will be in the drug component of healthcare.

Although the process systems engineering community has had some impact on the new drug discovery process, its most notable contributions have been to facilitate improvements in product development and manufacturing. Specifically these contributions have consisted of approaches for improving the design of processes for the production of the active ingredients, the design and development of formulated drug products and the robust operation of associated manufacturing processes. In recent years, the industry and the regulatory agencies have recognized the need for manufacturing innovation, one consequence of which is the accelerating transition from batch to continuous manufacturing. This in turn has driven changes in other components of the product life cycle, from process development to supply chain management. Moreover, this openness to change has stimulated further innovations in manufacturing methods such as 3D- and drop-on-demand printing of dosage, production of film-based products, such as roll-to-roll printing, as well as scaled-down, point of use manufacturing systems. PSE methodologies have important roles to play in all of these innovations.

Beyond these traditional product focused contributions, there also have emerged opportunities to deploy process systems methodology to facilitate the individualized administration of these products in real time or via regularly timed administration of specific dosage amounts. PSE modeling and optimization methodologies deployed in the form of automated closed loop or open-loop physician advisory systems can offer significant benefits to the patient, the clinician and the healthcare system. This bridging between the medicine and the clinical service provider points to further opportunities for PSE methodology in the healthcare services sector.

In this presentation, we will provide an overview of a selection of PSE contributions in these areas, drawing on work from the world-wide community, our collaborators and our own research group.

495589 Thermodynamic Simulations and Techno-Economic Analysis on the Utilization of CO_2 and a Novel Modularization Strategy for Chemical Looping Based Gtl Processes

Abstract:

The utilization of carbon dioxide (CO_2) as a feedstock in a chemical looping process that converts methane to syngas and a variety of downstream products has the ability to be impactful in the field of carbon capture, utilization, and sequestration (CCUS). The Ohio State University Chemical Looping methane to syngas (MTS) processâ€™ novelty comes from the coupling of carbon capture and utilization strategy, which captures and utilizes CO_2 to create a synergy atypical to many processes that are designed to deal with the challenges of carbon constraints. CO_2 utilization allows for increased flexibility of H_2:CO molar ratios that may be required for variations in the downstream processing while producing a comparable quantity of syngas with reduced natural gas input. In this study, thermodynamic models for chemical looping reactor systems are examined using ASPEN Plus. The chemical looping reducer reactor considered employs steam, natural gas and CO_2 as feedstock along with metal oxide oxygen carriers. This presentation will initially present system optimization based on thermodynamic Gibbs free energy minimization for the desired syngas composition required for downstream gas to liquid (GTL) processes. With CO_2 co-injection in the MTS process, the thermodynamic optimization indicates a natural gas savings of more than 20% over that of the baseline partial oxidation processes for GTL applications. Experimental verification using Thermo-gravimetric analyzer and bench-scale moving bed reactor that compared well with the thermodynamic model simulations will also be presented. A comprehensive techno-economic analysis on an energy, balanced (water balance, steam balance, heat and electricity balance) thermodynamic performance model of the MTS process with and without CO_2 utilization & modularization strategy was developed. The use of techno-economic analysis model for studying sensitivities associated with parameters including capital charge factor, first year versus leveled costs of fuel, crude oil equivalent prices and cost of chemical looping reactors will be presented. The comprehensive techno-economic analysis findings exemplify the importance of a CO_2 utilization and modularization strategy in chemical looping GTL processes and encourage further optimization in natural gas to commodity chemical conversion processes including gas to methanol, gas to ammonia, and gas to ethanol, amongst others.

534C:

Gasification of Residue from the Solvent Extraction of Powder River Basin Coal

William C. Schaffers
Ying Wang
David A. Bell (speaker)

American Institute of Chemical Engineers
National Meeting

November 1, 2017

UW UNIVERSITY OF WYOMING

Shenhua Direct Coal Liquefaction (DCL)

- Commercial plant in China that makes liquid fuels by hydrogenating coal.
- Residue from plant contains undigested coal + 1st stage catalyst
- This project:
  - Investigate feasibility of gasifying residue
    - Provide hydrogen for plant
    - Eliminate residue waste stream
  - Sponsored by U.S./China Clean Energy Research Center for Coal

UW

Project Assumptions

- Unable to contact Chinese team
- Assumed:
  - Shenhua DCL plant or similar built in U.S.
  - Powder River Basin coal feed.
  - Need to recommend gasification technology for residue.

Fluidized bed gasifiers

- Lower cost than entrained flow gasifiers
- Requires reactive feedstock
  - Biomass
  - Lignite
  - Sub-bituminous coal, such as PRB
- Bituminous coals require more severe conditions in entrained flow gasifier.

Experimental strategy

- Approximate coal liquefaction process by extraction with tetralin, a hydrogen donor solvent

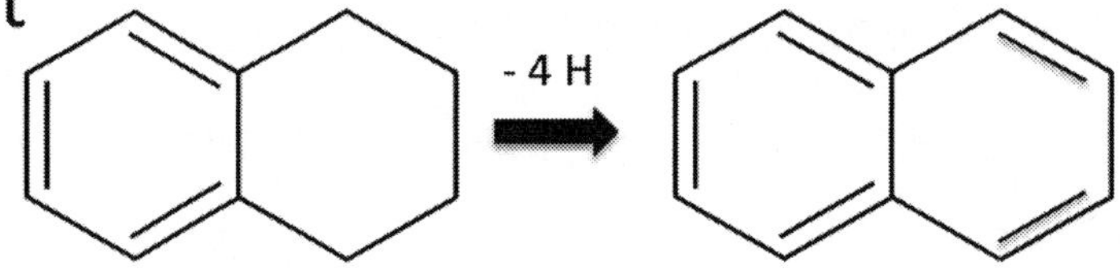

- Compare gasification reactivity of extraction residue to parent coal.

UW

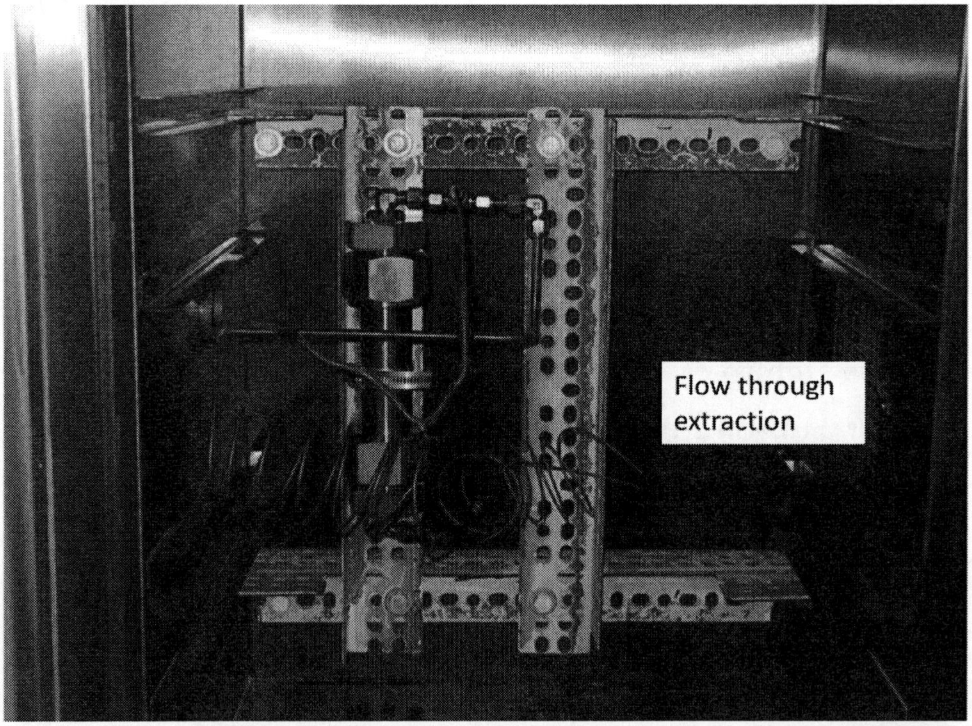

UW

UW

Extraction at 360 C

Run #	As Received Weight (g)	Dry Weight (g)	Moisture Loss (wt %)	Extraction Residue Weight (g)	Material Extracted (wt %)
3	54.38	46.54	14.42	25.31	45.62
5	57.88	50.19	13.29	30.01	40.21
6	60.27	51.38	14.75	28.39	44.75

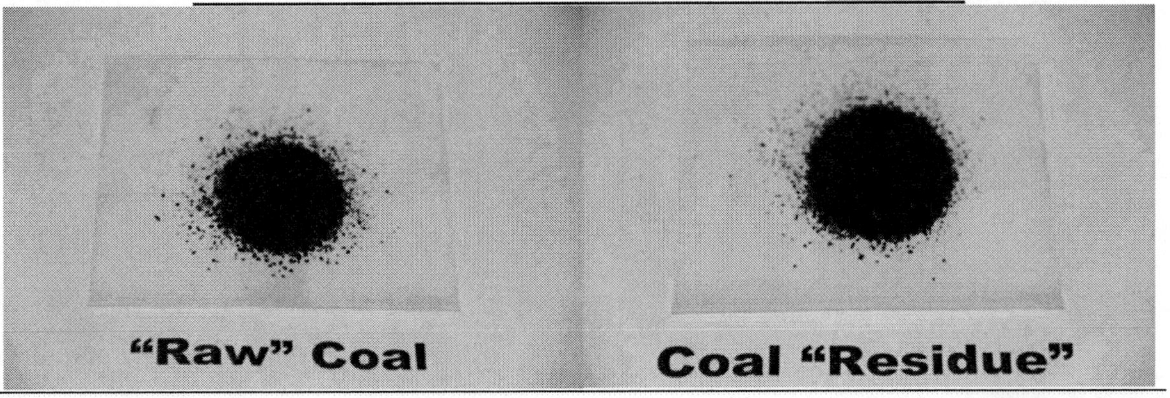

UW

Proximate & Ultimate Analyses

	As Received Basis (wt %)		Moisture Free Basis (wt %)		Moisture & Ash Free Basis (wt %)	
	Decker Coal	*Extraction Residue*	*Decker Coal*	*Extraction Residue*	*Decker Coal*	*Extraction Residue*
Proximate Analysis (Method: ASTM D-5142; D-7582)						
Moisture	24.29	*3.27*	****	****	****	****
Ash	3.62	*9.78*	4.78	*10.11*	****	****
Volatile Matter	29.42	*31.48*	38.86	*32.54*	40.81	*36.20*
Fixed Carbon	42.67	*55.47*	56.36	*57.35*	59.19	*63.80*
Total	**100.00**	*100.00*	**100.00**	*100.00*	**100.00**	*100.00*
Ultimate Analysis (Method: ASTM D-5142/5373; D-7582/4239)						
Moisture	24.29	*3.27*	****	****	****	****
Hydrogen	2.88	*3.80*	3.80	*3.93*	3.99	*4.38*
Carbon	56.48	*68.28*	74.59	*70.59*	78.33	*78.53*
Nitrogen	0.87	*1.05*	1.15	*1.09*	1.21	*1.21*
Sulfur	0.34	*0.28*	0.45	*0.29*	0.47	*0.32*
Oxygen	11.52	*13.54*	15.23	*14.00*	16.00	*15.57*
Ash	3.62	*9.78*	4.78	*10.11*	****	****
Total	**100.00**	*100.00*	**100.00**	*100.00*	**100.00**	*100.00*

UW

Gasification tests

- Modified drop tube reactor
- Preheated, gas flow established
- ~1 g of coal or residue dropped into reactor
- Lands on glass frit in center
- Effluent gases measured using mass spectrometer

UW

Typical mass spectrometer results

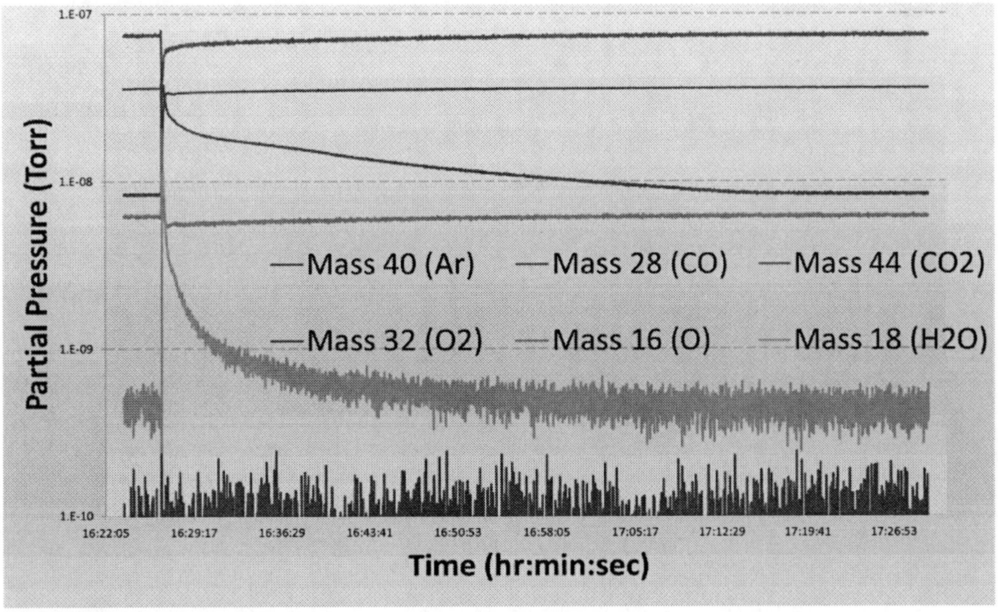

UW

Data fit

- Reaction complete when mass spectrometer signal returns to baseline
- Integration of signal versus time used to measure extent of reaction, *X*.
- Data fit to random pore model (assumes reaction on pore walls):

$$dX/_{dt} = \boldsymbol{k_p}(1 - X)\sqrt{1 - \psi ln(1 - X)}$$

Where $\boldsymbol{k_p}$ is the rate parameter, ψ is dimensionless structural parameter.

UW

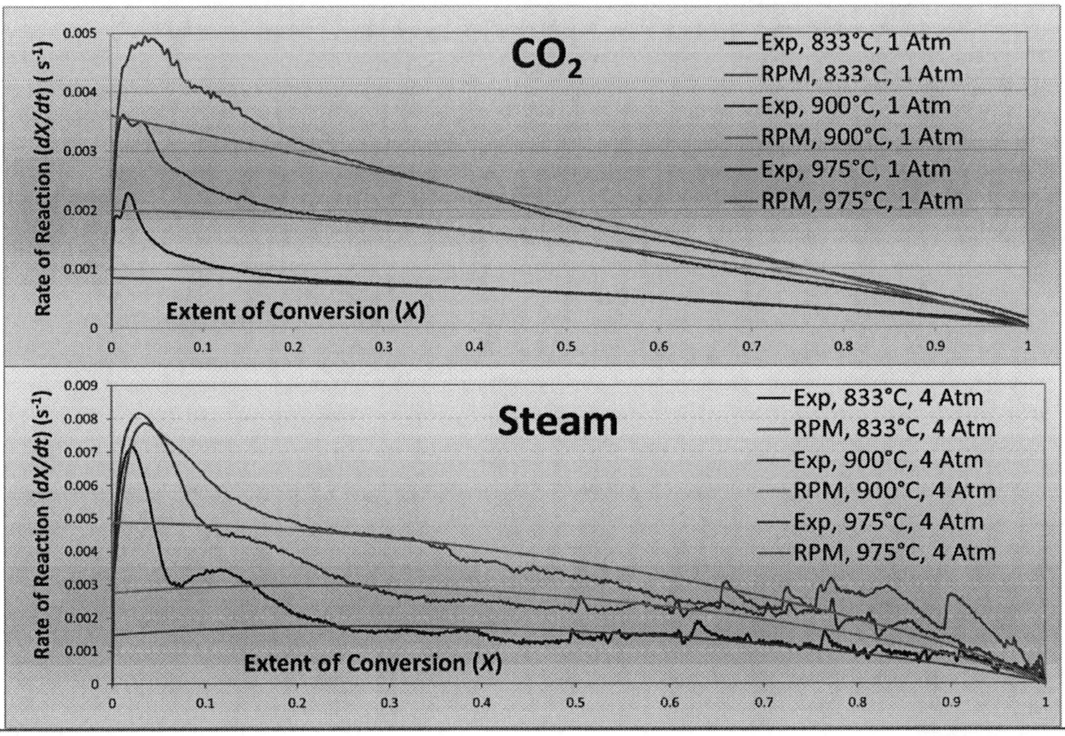

Coal & Residue Gasification in CO_2

Pressure	Temp.	Coal Time (hr:min:s)	Residue Time (hr:min:s)	*Coal* $k_p (s^{-1})$	*Residue* $k_p (s^{-1})$	*Coal* ψ	*Residue* ψ
1 atm	833 °C	1:06:12	0:47:45	4.10E-04	8.75E-04	3.37	1.50
	900 °C	0:25:00	0:22:52	1.20E-03	1.81E-03	3.12	1.51
	975 °C	0:12:06	0:15:28	2.75E-03	3.44E-03	2.45	0.63
4 atm	833 °C	1:01:36	1:00:02	5.00E-04	8.17E-04	2.65	0.92
	900 °C	0:23:00	0:28:36	1.28E-03	1.95E-03	2.63	0.28
	975 °C	0:11:00	0:17:11	2.50E-03	3.88E-03	3.25	0.10

Time is time to complete gasification.

Residue gasification is usually faster than gasifying parent coal.

Coal & Residue Gasification in Steam

Pressure	Temp.	Coal Time (min:s)	Residue Time (min:s)	*Coal* $k_p (s^{-1})$	*Residue* $k_p (s^{-1})$	*Coal* ψ	*Residue* ψ
1 atm	833 °C	17:41	13:19	1.28E-04	2.01E-03	5.25	4.69
	900 °C	9:26	9:34	2.75E-03	3.95E-03	5.25	2.18
	975 °C	4:31	7:07	6.30E-03	4.67E-03	3.50	3.39
4 atm	833 °C	19:53	19:19	1.28E-04	1.43E-03	3.63	4.27
	900 °C	13:15	10:02	2.60E-03	3.00E-03	2.95	3.65
	975 °C	6:47	07:31	5.85E-03	4.89E-03	2.23	2.88

Steam gasification faster than CO_2 gasification.

Residue gasification usually faster than coal gasification.

Surface Areas of Chars from Coal & Residue

Method	Surface Area (m^2/g)					
	Coal Char (833°C)	Residue Char (833°C)	Coal Char (900°C)	Residue Char (900°C)	Coal Char (975°C)	Residue Char (975°C)
BET (N_2) (Mesopores)	0.36	10.81	1.49	24.39	34.54	40.25
D-R (CO_2) (Micropores)	408.51	382.13	341.44	382.61	344.68	404.13

Calcium and Sodium on Char Particles

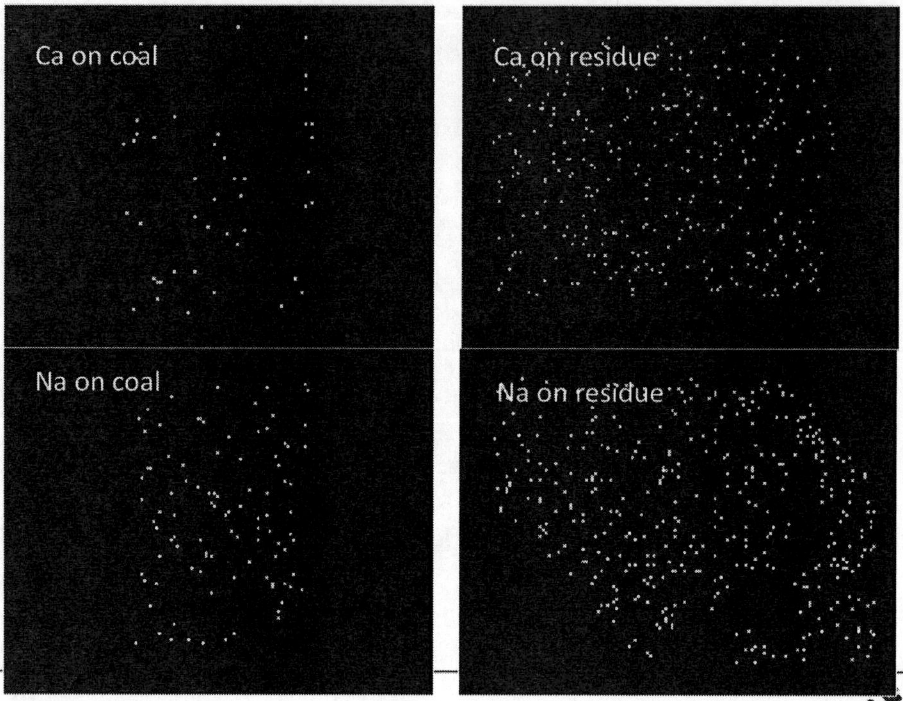

UW

Conclusions

- Residue is more reactive than parent coal
- Effect may be due to
  - Increase in mesoporous surface area
  - Concentration of naturally-occurring catalysts, sodium and calcium
- Residue can be gasified in fluidized bed gasifier.

UW

Development of a Modular Coal/Biomass to Liquids Research Facility

DOE Project Number: FE0010482

Andy Placido

University of Kentucky, Center for Applied Energy Research
http://www.caer.uky.edu/powergen/home.shtml

Facility Overview

- Integrated coal/biomass-to-liquids (CBTL) facility at a capacity of 1 bbl./day
- Purposely designed for maximum flexibility: modular, skid-mounted, anticipating frequent change-outs; "plug and play;" and future re-purposing

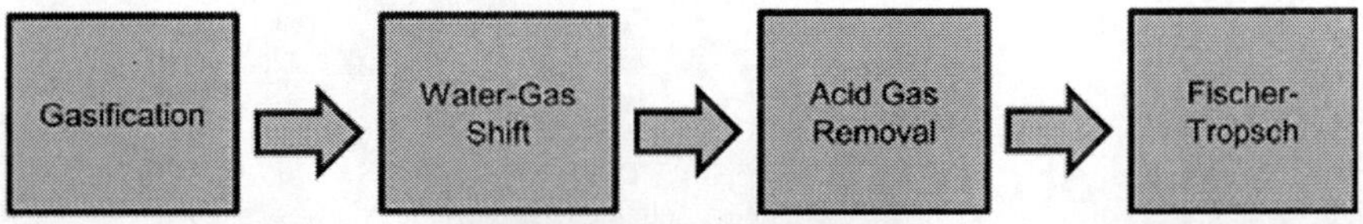

Center for Applied Energy Research

caer.uky.edu

OMB Gasification System

- Entrained flow gasifier
  - Coal/water slurry
  - Oxygen blown
  - Molten slag
  - 1 ton/day coal consumption
- Includes biomass/coal feed preparation tower
  - Hydro-vibrating Ball mill

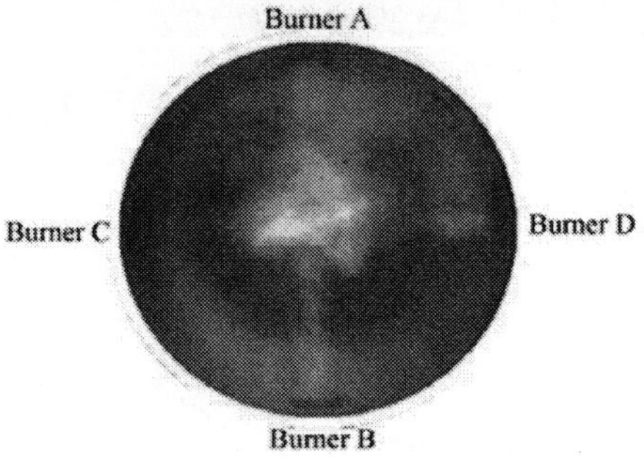

2017 AIChE Annual Meeting

November 1st, 2017

Center for Applied Energy Research

caer.uky.edu

Feed Preparation

- Ash fusion temperature generally limiting factor of feed
- Capable of utilizing wide range of coals and blends
- Designed to handle torrefied biomass
- Important Final Slurry Properties
  - Particle size
  - Viscosity

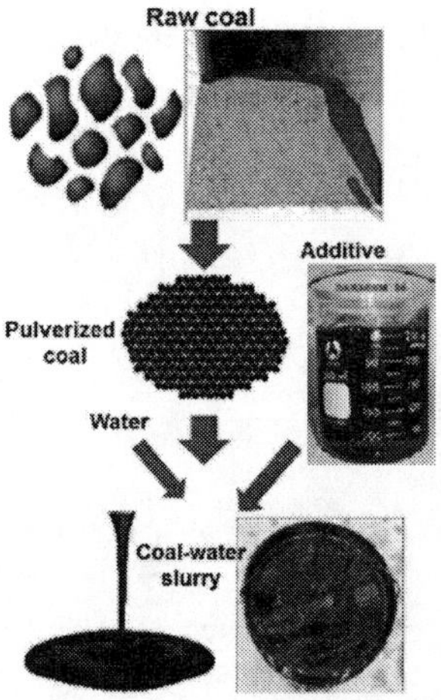

Typical Properties of CWS

Average particle size (μm)	Mass concentration	Viscosity (mPa·s)
<50	<60%	<250

2017 AIChE Annual Meeting

November 1st, 2017

Center for Applied Energy Research

caer.uky.edu

WGS

$CO + H_2O \Leftrightarrow CO_2 + H_2$ Tunable H_2/CO ratio

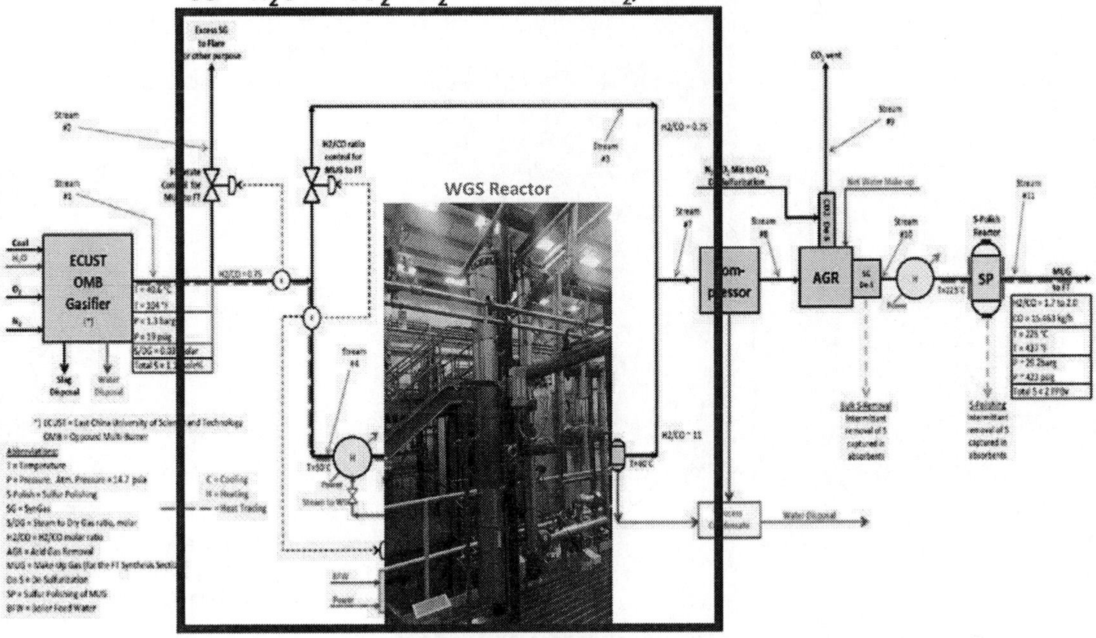

2017 AIChE Annual Meeting November 1st, 2017

Center for Applied Energy Research

caer.uky.edu

Acid Gas Removal

Aqueous Solvent System

- CAER has 10+ years Experience with Similar Technology
- Thermal and Pressure Swing Absorption
- Howden Burton Corblin metal diaphragm compressor
  - Pressures up to 450 psi
- Wide range of Amine Solvents
- Sulfur Treatment on Both Rejected Acid Gas and Treated Gas
  - Activated carbon
- 95+% removal of CO_2
- 99.99% removal of sulfur (less than 1ppm)

2017 AIChE Annual Meeting November 1st, 2017

Center for Applied Energy Research
caer.uky.edu

Downstream - FT

- Chart Energy microchannel reactor with production capacity of 1bbl./day
- FT Reactor is aluminum heat exchanger inside a steel kettle
- Capable of utilizing either Fe or Co based catalysts
  - Difference in H_2/CO molar ratio
- Wide ranging fuel production
- Water Gas Shift and FT modules were designed and built by Zeton Inc. with assistance from Stovlbaek Consulting

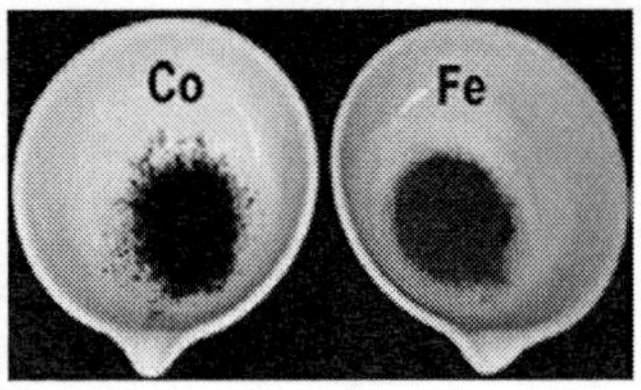

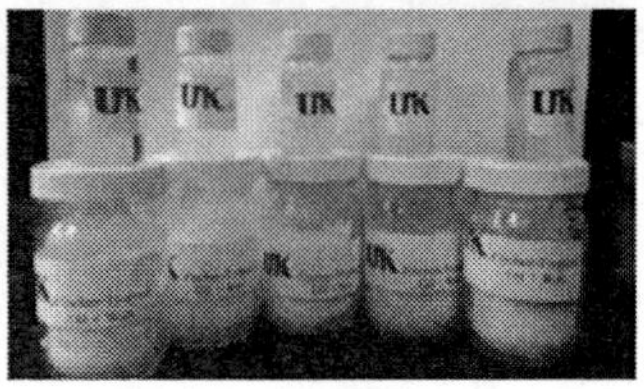

2017 AIChE Annual Meeting

November 1st, 2017

Center for Applied Energy Research
caer.uky.edu

Balance of Plant

2017 AIChE Annual Meeting

November 1st, 2017

Center for Applied Energy Research

caer.uky.edu

Experimental Data from Coal Only

- So far Gasifier test operation only
- Typical Operation data for 24 hour period
- Operationally stable for entire testing run
- Feed details
  - 30-40% coal/60-70% water
  - Co-fire small amount of natural gas (~15% heat input) to maintain heat

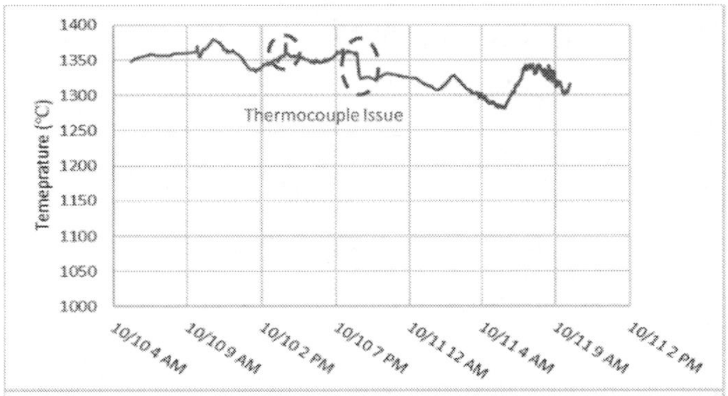

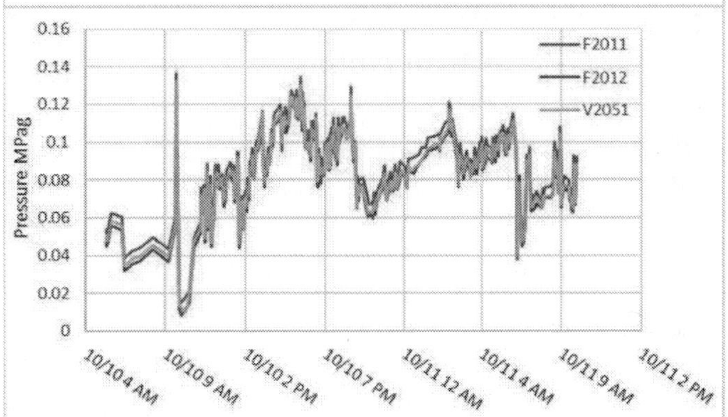

Center for Applied Energy Research

caer.uky.edu

Experimental Data from Gasifier Operation

- Overall availability ~50%
  - First operation for syngas quality control testing
  - GC learning curve (filters clogging, learning software and hardware)
- H_2/CO important syngas quality measure for downstream
  - Ratio stable around 1.1
  - Higher than expected due to co-firing with natural gas

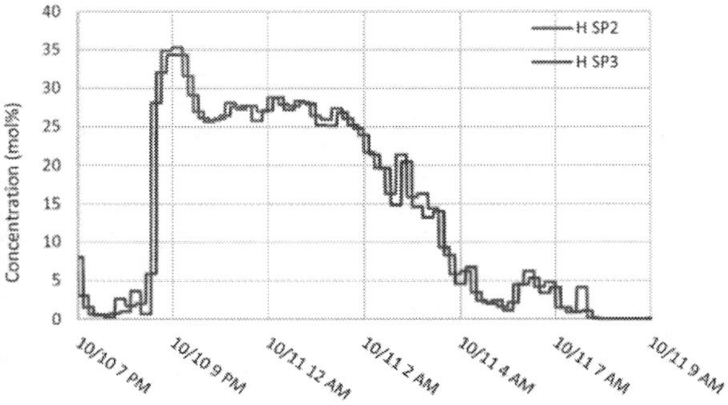

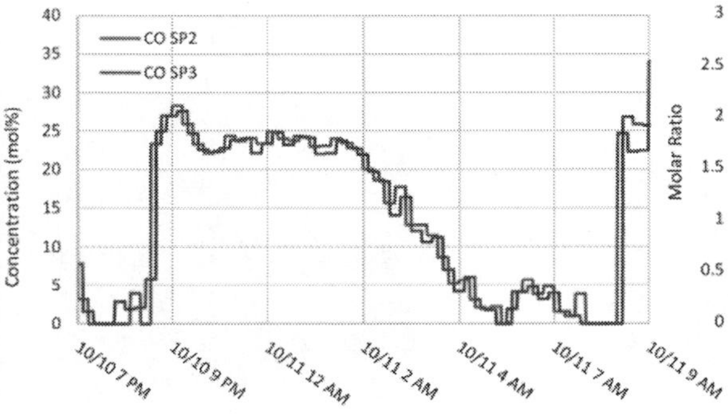

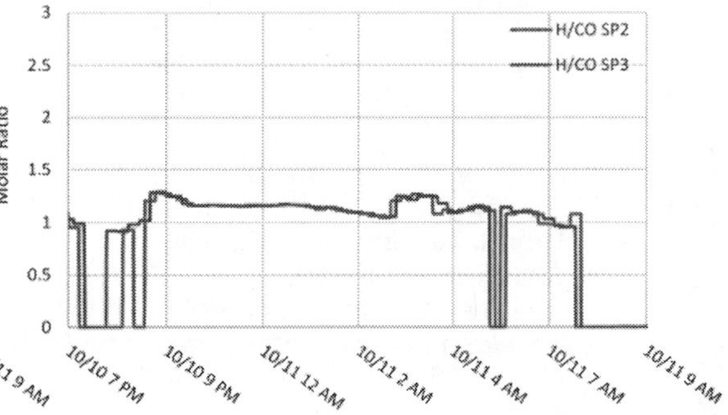

Experimental Data from Gasifier Operation

- Coal Conversion
  - CO and CO_2 produced
  - Total carbon fed (coal and natural gas)
- Low Conversion
  - First trial run
- Burner Atomization
- Slurry Concentration

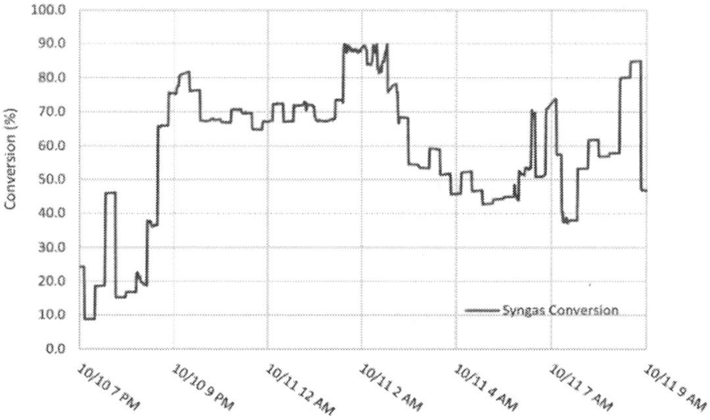

Lessons Learned

- Finding pilot scale process equipment within budget and scope often require exhaustive and creative search
- The Project Benefited Greatly from Strong Collaborations in China in the area of Coal Gasification while also posing some unique challenges
- Slurry concentration has a dramatic affect on gasifier operation and gas composition
  - Challenge to maintain constant concentration (density)
  - Blend based on density and determining acceptable density range limits
  - Additive for settling and pumping
- Refractory on this small scale has caused some integrity issues with thermocouples
  - Thermal expansion/contraction
  - Wrong initial design
  - New thermocouples to meet our gasifier needs
  - Development/utilization of temperature probes without physical insertion (IR, Pyrometer, etc)

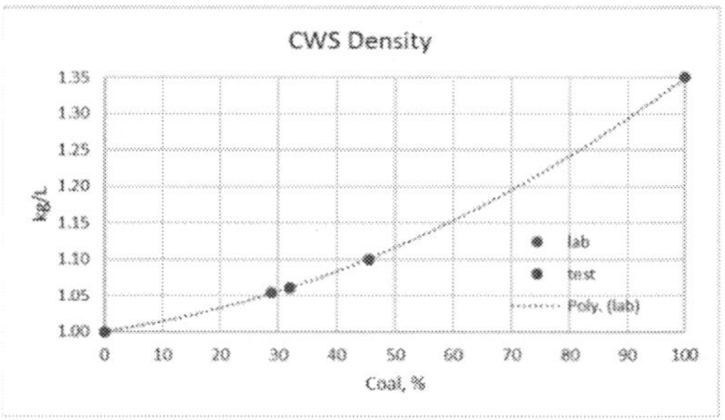

Gasifier Thermocouple Well

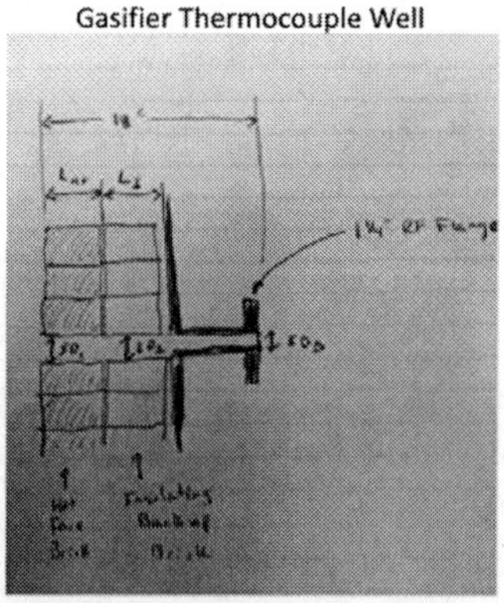

Summary

- Modular nature of the facility provides wide range of capabilities and flexibility for future use
- Whole facility ready for operation
- Facility is available for testing

2017 AIChE Annual Meeting November 1st, 2017

Acknowledgements

- DOE-NETL: Arun Bose, Jenny Tennant and Jason Hissam
- UK: Rodney Andrews, Don Challman, Kunlei Liu, Burt Davis, Dennis Sparks, Steve Summers, Tate VanHoose, Len Goodpaster, Otto Hoffman, Marshall Marcum, Brad Irvin, Jimin Zeng and Jonathan Pelgen
- Zeton: Ravin Dave, Danielle De Sousa, Erik Gaspar
- Stovlbaek Consulting: Peter Pedersen
- ECUST: Qinghua Guo, Yan Gong, Jianliang Xu
- CEEDI: Wang Qiuming and Tao Li

2017 AIChE Annual Meeting November 1st, 2017

507270 Production of Fuels from Coal-Biomass Feedstock using Fischer-Tropsch

Abstract:

The Fischer-Tropsch synthesis is actually composed of four sub-processes: gasification, gas clean-up, Fischer-Tropsch synthesis and product upgrading. The impact of feedstock (coal, natural gas, biomass) and mixtures of these feedstocks upon each sub-process will be discussed. The major impact of the feedstock is on the first two sub-processes: gasification and gas clean-up.

499845 Coal Power Plants with Enhanced Profitability and No Carbon Dioxide Emissions

Abstract:

Coal Power Plants with Enhanced Profitability and No Carbon Dioxide Emissions

A zero carbon dioxide emission process is developed, in which electrical power and formic acid are coproduced from coal. The overall reaction for the proposed process uses coal, oxygen, and water as reactants to generate formic acid as the only product. A thermodynamic analysis is carried out, so that feasible operating limits for the proposed process are identified. Since the process cannot be realized within one single step, a reaction cluster is proposed which consists of feasible chemical reactions. These are: gasification of coal to syngas; CO_2 hydrogenation to formic acid; hydrogen combustion; and CO combustion. The principal technologies employed in this process network include combustion, gasification of coal to syngas, gas compression, supercritical fluid pumping, hydrogen separation using pressure swing adsorption, and formic acid production from CO_2 and H_2. The software UniSim [TM, Honeywell Inc] is employed to simulate the flowsheet. Two Gibbs reactors act as the gasification reactor and a combustor, while a conversion reactor carries out the formic acid reaction. Coal, water, and part of oxygen are fed into the gasification reactor; off-gas from the hydrogen PSA and oxygen are fed into the combustor; pure hydrogen from hydrogen PSA and CO_2 which is compressed to supercritical phase are fed into the formic acid reactor. By adjusting the oxygen delivery ratio, the coal gasification is designed to be energetically self-sufficient, so that the whole process produces the largest amount of power. Heat exchangers, phase separators, compressors, pumps, etc. are also employed to realize the process. Temperature and heat load information for each process stream is obtained from the software simulation and used in carrying out a heat and power integration study. Through the use of heat engine and heat pump subnetworks, a feasible operating envelope for the considered process is identified. A preliminary technicoeconomic analysis is carried out, which establishes the superiority of the proposed process to current coal power plant technology.

501636 Comparative Study of a Hybrid Adsorptive-Membrane Reactor (HAMR) with a Membrane Reactor/Adsorptive Reactor Sequence

Abstract:

Reactive separation processes such as membrane reactors (MRs) and absorptive reactors (ARs) have been attracting renewed interest for industrial applications. We have been studying a novel process that combines the MR and AR functions and which shows good promise for application in the efficient ultra-pure hydrogen production with simultaneous CO_2 recovery and capture.. Two system configurations are being investigated: One (known as the Hybrid Adsorptive-Membrane Reactor or HAMR) in which the membrane, adsorbent and catalyst are all housed in the same unit; and another in which the MR and the AR are separate but coupled units (known as the MR-AR system). The aim of this study is to perform a comparative numerical study between the HAMR and the combined MR-AR system, in which the MR is followed by an adsorptive reactor (AR), for the case-study of the water gas shift (WGS) reaction process for hydrogen production and carbon dioxide capture.

The HAMR is a dynamically-operated process, which typically employs one or multiple tubular membranes. The HAMR's reaction section, surrounding the membrane tubes, contains catalyst and adsorbent material that accelerates reaction kinetics via the adsorption of one (or several) of the reaction products. The HAMR's permeation section consists of the membrane tubes which physically remove some of the other reaction products from the reacting mixture, further accelerating reaction kinetics. The HAMR is, typically, operated in reaction and regeneration modes, since the adsorbent meterial reaches its adsorption capacity after a period of operation, and thus needs to be regenerated.

The MR-AR configuration (with the AR following the MR) provides significant added flexibility for the application, for which in addition to efficient hydrogen production CO_2 recovery and purity are also important key drivers. In the MR, a membrane that is selective to hydrogen is used to enhance the WGS reaction rate, and to potentially overcome equilibrium conversion limitations imposed by thermodynamics. The MR system is composed of a reaction zone packed with catalyst pellets, and a permeation zone, where the reaction products permeate. The AR contains both catalyst and adsorbent in a fixed-bed configuration for simultaneous reaction and separation. The combined MR-AR as with the HAMR is a dynamically operated process.

In the study, the velocity and species concentration profiles along the reactors' lengths are captured by momentum/species transport models accounting for convection/reaction /diffusion mechanisms. The model's equations are solved using Finite Element Method (FEM). The rigorous Maxwell-Stefan and dusty gas models are applied to describe mass diffusion fluxes. The developed model is used to intensify the Water Gas Shift Reactor (WGSR) Process. Then parametric studies of the HAMR and combined system are carried out, so as to identify maximum intensification designs. These studies include a broad range of operating conditions and parameters (e.g., reactor operating temperature, catalyst and adsorbent weight to feed flowrate ratios, and others).

493317 Zeolite Membrane Reactor for High-Temperature Isobutane Dehydrogenation Reaction: Effects of Membrane Properties and Operating Conditions

Abstract:

Zeolite Membrane Reactor for High-Temperature Isobutane Dehydrogenation Reaction:

Effects of Membrane Properties and Operating Conditions

*Shailesh Dangwal, Ruochen Liu, Seok-Jhin Kim**

School of Chemical Engineering, Oklahoma State University,

420 Engineering North, Stillwater, OK 74078, USA

ABSTRACT

Isobutylene plays an instrumental role as an intermediate in many important chemical engineering processes. Its use in making gasoline oxygenates, fuel additives and butyl rubber adds up to its demand and stimulates research for new energy efficient ways for its production. Catalytic dehydrogenation of isobutane using a membrane reactor is a feasible solution not only because it helps in overcoming thermodynamic limitations by shifting the equilibrium towards isobutylene but also it substantially separates isobutylene by preferentially allowing only H_2 to permeate through the membrane. In this study, dehydrogenation of isobutane in packed-bed membrane reactors operating with a Pt/Al_2O_3 catalyst is reported. The effects of MFI zeolite membranes on isobutane conversion, isobutylene selectivity and isobutylene yield are investigated experimentally and by simulations using a simple one-dimensional model. MFI membrane reactors allow the thermodynamic equilibrium limit of isobutane conversion to be surpassed at high temperatures. The experimental results and the model calculations demonstrate that medium-pore MFI membranes with good $H_2/i\text{-}C_4H_{10}$ selectivity can effectively enhance isobutane conversion at high operation temperature by timely removal of H_2 through the membranes. *Corresponding author: E-mail: seokjhin.kim@okstate.edu

499695 Ideas As a Process Intensification Tool Applied to Process Networks Containing Membrane Reactors

Abstract:

Process intensification is a strategy that can make dramatic reductions (order 100 or more) in the size of a chemical plant while meeting production objectives[1]. More recently, process intensification can be defined as "any chemical engineering development that leads to substantially smaller, cleaner, and more energy efficient technology[2]". In this work, the Infinite-DimEnsionAl State-space (IDEAS) conceptual framework is employed as a process intensification activity applied to process networks containing membrane reactors. The IDEAS framework enables a broad search of the design state-space to identify optimal process networks through the formulation of linear programs. This framework can assess fundamental performance limitations of networks of technologies under consideration in order to identify intensified designs. The IDEAS methodology for membrane containing networks will be applied to a case study featuring natural gas based hydrogen production, where energy efficiency will be intensified.

Membrane reactors are an emerging technology in the steam-methane reforming (SMR) process; these reactors typically involve simultaneous reaction and separation[3]. The biggest advantage of this technology is that it allows in situ separation, which gives rise to a high-purity product. This is particularly useful in driving the forward reaction of equilibrium limited reactions, as product is continuously removed from the reactor. The use of equilibrium reactor models allows for the search of intensified designs to be carried out in the lower-dimensional space of atomic fraction ratios, rather than the space of species mole fractions.

The process network considered will feature pressure changing devices, equilibrium reactors, a heat exchange network, flash separators, ideal separators, and membrane reactors. Operating and capital cost studies are all presented to determine the process network's cost relative to traditional steam methane reforming.

1. Ramshaw, C. The incentive for process intensification. BHR Group Conference Series Publication, vol. 18, pp. 1-4. Mechanical Engineering Publications Limited, 1995.
2. Stankiewicz AI, Moulijn JA. Process intensification: transforming chemical engineering. Chemical Engineering Progress. 2000;96(1):22-34.
3. Soltani, S., Sahimi, M., and Tsotsis T.T., "Catalytic Membrane Reactors: A Brief Overview", Encyclopedia of Membrane Science and Technology, Eric. M.V. Hoek and Volodymyr V. Talabar Editors, John Wiley & Sons, 2013.

500575 Additive Manufacturing for Process Intensification: Tailor-Made Design of Catalyst Supports for Single Phase and Multiphase Reaction Systems

Abstract:

Open cellular structures are promising novel catalyst supports as they combine the advantages of randomly packed fixed-bed reactors (radial mixing, tortuosity of the flow) and honeycombs (high geometric specific surface area, low pressure drop) owing to their high porosities and their characteristic 3D cellular architecture. Open cellular structures are manufactured mainly by means of a replication technique, which results in an irregular architecture of the structure, and consequently of the open porous system.

With the rise of additive manufacturing techniques it is nowadays possible to manufacture periodic open cellular structures (POCS) by, e.g., selective electron beam melting (SEBM) with well-defined geometrical properties. In this regard, a new degree of freedom in the design is unlocked as metal-based open cellular structures of nearly arbitrary geometrical complexity can be manufactured in a well-defined and highly reproducible manner.

We have extensively investigated open cellular materials with both periodic [1,2] and random [3] cell geometries regarding their performance as catalyst support. These structures were characterized with respect to their morphological/geometrical properties, fluid dynamic behavior and mass and heat transport properties. Models for the estimation of properties important for reactor design such as, e.g., specific surface area and pressure drop [1, 2] were developed and validated by experiments.

Detailed numerical studies of such geometries and the 3D flow field along with mass and heat transport processes as well as studies on the effective heat conductivity in POCS of different configurations were carried out [3, 4]. The established models allow for the design and optimization of POCS that are tailor-made according to the needs of the reaction system, representing a new class of superior catalyst supports.

The tailor-made POCS were manufactured, functionalized by catalytic coating, and then applied in two case studies for demonstration. In the first case study, the focus is on the optimization of heat transport in a highly exothermic gas phase reaction for enhancing the selectivity, where the oxidation of methanol to formaldehyde serves as example system. In the second case study the phase contacting, i.e. the gas-liquid distribution in a trickle-bed reactor was optimized [5]. As example reaction system, the hydrodesulfurization (HDS) of gas oil was investigated.

For both reaction systems, the tailor-made POCS were benchmarked against a traditional randomly packed bed reactor, and clearly a great potential for process intensification could be demonstrated.

References

[1] Inayat, A.; Schwerdtfeger, J.; Freund, H.; Koerner, C.; Singer, R.F.; Schwieger, W.; Chem. Eng. Sci. 66(12) (2011) 2758-2763.

[2] Klumpp, M.; Inayat, A.; Schwerdtfeger, J.; Koerner, C.; Singer, R.F.; Freund, H.; Schwieger, W.; Chem. Eng. J. 242 (2014) 364-378.

[3] Bianchi, E.; Groppi, G.; Schwieger, W.; Tronconi, E.; Freund, H.; Chem. Eng. J. 264 (2015) 268-279.

[4] Bianchi, E.; Schwieger, W.; Freund, H.; Adv. Eng. Mater. 18(4) (2016) 608-614.

[5] Laemmermann, M.; Schwieger, W.; Freund, H.; Catal. Today 273 (2016) 161-171.

485501 Light Absorption Efficiency of Gas-Liquid Continuous Flow Microreactors Illuminated By Visible LED Light Sources

Abstract:

The use of light energy constitutes a green activation mode of organic molecules. A photoreaction of increasing interest is the photosensitized addition of singlet oxygen, as it is an atom-economic method of functionalization and is applied in the synthesis of commercial chemical products as fragrances and pharmaceuticals [1].

Continuous photo microreactors are a promising tool for e.g. photooxidation transformations due to their small penetration depth and the presence of segmented flow which leads to enhanced mass transfer of the gas in the liquid [2]. In addition, the large photon flux through the thin liquid film located between the bubble and the reactor wall and light scattering at the gas-liquid interface were also reported as possible factors to enhance the photoreaction yield in comparison to single phase flow [3, 4]. However, designing an efficient photoreactor does not only involve optimizing its geometry, but also the light source needs to be adapted to the chosen chemistry, reactor channel layout and the phase of the reaction medium [5, 6]. In recent years, the use of LEDs has become popular in flow photochemistry due to their energy efficiency and the possibility to match their emitted wavelength to the photosensitizer absorption and their size to the reactor microchannels. However, the influence of the LED light distribution and potential scattering of light at the gas-liquid interface on the yield of the photochemical reactions performed in segmented flow is so far not reported.

This study aims to investigate the influence of the LED board configuration on the photon absorption efficiency in gas-liquid continuous flow microreactors. Firstly, the distribution and number of photons emitted by LED boards with different configurations is experimentally determined by near field goniophotometric measurements (NFGM). Secondly, the number of photons absorbed in the microreactor channels is quantified using an actinometer system. The balance of the photon fluxes for various LED light sources and for operating in single phase liquid or two-phase gas-liquid flow are compared and discussed. Furthermore, we developed a modelling tool based on the ray tracing technique to simulate the distribution of the scattered light rays from the gas-liquid interface, which supports the discussion of the experimental observations.

This study will contribute to the understanding of the incident radiation and dispersed phase influence on the photon absorption in gas-liquid photochemical transformations, knowledge which will be crucial for the efficient design of multiphase flow photochemical systems.

References

1. J. P. Knowles, L. D. Elliott, and K. I. Booker-Milburn, Flow photochemistry: Old light through new windows. Beilstein Journal of Organic Chemistry, 2012. 8: p. 2025-2052.

2. A. Gavriilidis, A. Constantinou, K. Hellgardt, K. K. (Mimi) Hii, G. J. Hutchings, G. L. Brett, S. Kuhn, and S. P. Marsden, Aerobic oxidations in flow: opportunities for the fine chemicals and pharmaceuticals industries, Reaction Chemistry & Engineering, 2016. 1: p. 595-612.

3. A. Yavorskyy, O. Shvydkiv, C. Limburg, K. Nolan, Y. M. C. Delauré, and M. Oelgemöller, Photooxygenations in a bubble column reactor, Green Chemistry, 2012. **14**: p. 888-892.

4. M. Nakano, Y. Nishiyama, H. Tanimoto, T. Morimoto, and K. Kakiuchi, Remarkable Improvement of Organic Photoreaction Efficiency in the Flow Microreactor by the Slug Flow Condition Using Water, Organic Process Research and Development, 2016. 20: p. 1626–1632.

5. M. E. Leblebici, G. D. Stefanidis, T. Van Gerven, Comparison of photocatalytic space-time yields of 12 reactor designs for wastewater treatment, Chemical Engineering and Processing, 2015. 97: p. 106-111.

6. D. Ziegenbalg, G. Kreisel, D. Weiß, and D. Kralisch, OLEDs as prospective light sources for microstructured photoreactors, Photochemical & Photobiological Sciences, 2014. 13: p. 1005-1015.

502576 Fast Mixing, Ketimine Additions, and the Application of Flow Chemistry to Verubecestat

Abstract:

Alzheimerâ€™s disease (AD) is an epidemic of ever increasing concern that affects over 24 million patients worldwide. Research suggests that the formation of Î²-amyloid plaque is central to the onset of AD and that inhibition of the amyloid pathway could lead to a disease modifying therapy. One of the first steps in the formation of plaque involves the enzyme Î²-secretase (BACE-1). Clinical programs for BACE inhibitors are progressing at a number of companies including MSD where Verubecestat (MK-8931) is in Phase III.

During the commercial route development of MK-8931, it was determined that the nucleophilic addition to an imine could be improved by performing the reaction in flow. The enhanced mixing lead to an increase in the yield of the reaction while excellent heat transfer properties of a flow reactor lead to the elimination of the need for cryogenic conditions.

The development of this process from an initial lab scale hit through a successful demonstration in the prep lab at multi kilogram scale to a pilot plant campaign on hundreds of kilograms is presented.

Synthesizing high selectivity bromobutyl rubber by a microreactor system

Pei Xie, Guangsheng Luo*, Kai Wang*, Peijian Wang

The State Key Laboratory of Chemical Engineering, Department of Chemical Engineering, Tsinghua University, Beijing, 100084, China

xie-p14@mails.tsinghua.edu.cn, 86-10-62797493

Skloche

Outline

- **Background**
- **Experimental section**
- **Results and discussion**
- **Conclusions**

Bromobutyl rubber (BIIR)

High performance properties :

- Low-permeability
- Ageing resistance
- Ozone resistance

Δ High rautohension, mutual viscosity

Δ Excellent compatibility with other unsaturated rubber

Δ Fast vulcanizing speed

Wildly used in:

- Tubeless tyre
- Medical rubber products
- Rubber hoses
- …

Principle

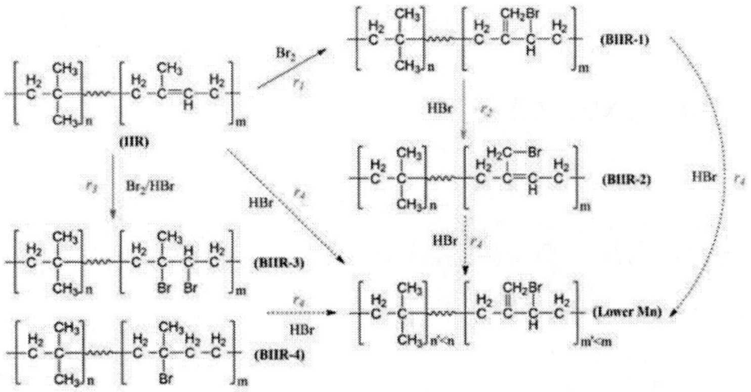

☐ Characteristics of this system:

- High viscosity } Mixing-limited
- Extreme flow rate ratio of IIR solution to Br_2 solution } Mixing-limited
- Low selectivity of main product — Shorten resistence time & eliminate HBr quickly
- Strong corrosion from Br_2 and HBr — Develop a reactor made by corrosion resiatant material

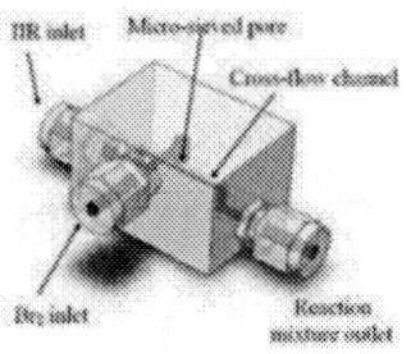

Experimental set-up

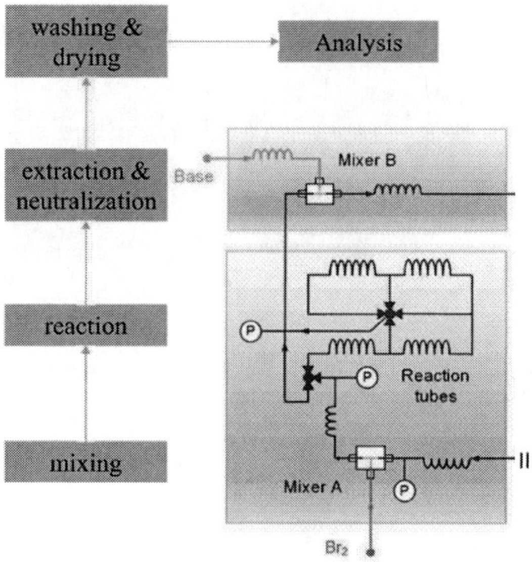

□ NMR analysis

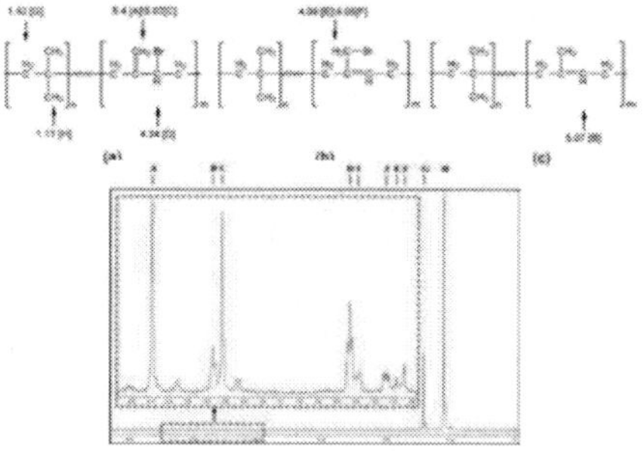

$$X = \frac{m_{eff\text{-}Br}}{m_{BIIR}} = \frac{\left([D]+\frac{[E]+[F]}{2}\right)\cdot M_{Br}}{\left([D]+\frac{[E]+[F]}{2}\right)(M_{C5}-M_{H}+M_{Br})+[B]\cdot M_{C5}+\frac{[H]}{6}\times M_{C4}}$$

$$w = \frac{n_{BIIR\text{-}1}}{n_{BIIR\text{-}1}+n_{BIIR\text{-}2}} = \frac{[D]}{[D]+\frac{[E]+[F]}{2}}$$

$$U = \frac{\sum n_{\text{double-bond unit}}}{\sum n_{\text{double-bond unit}}+n_{C4}} = \frac{[B]+[D]+\frac{[E]+[F]}{2}}{[B]+[D]+\frac{[E]+[F]}{2}+\frac{[H]}{6}}$$

□ GPC analysis

- *Mn* (number average molecular weight)
- *PDI* (polydispersity index)

Outline

- **Background**
- **Experimental section**
- **Results and discussion**
- **Conclusions**

Results and discussion

☐ Influence of residence time

x_{IIR} (wt%)	x_{Br2} (wt%)	x_{NaOH} (wt%)	$x_{Na2S2O5}$ (wt%)
10.4	5.8	0.5	0.1
Q_{IIR} (mL/min)	Q_{Br2} (mL/min)	Q_{base} (mL/min)	T (°C)
90.3	9.4	49.7	25

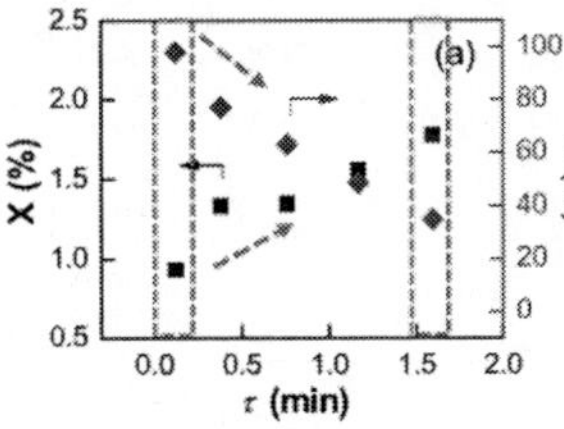

- HBr could catalyze the Br substitution position transfer reaction.

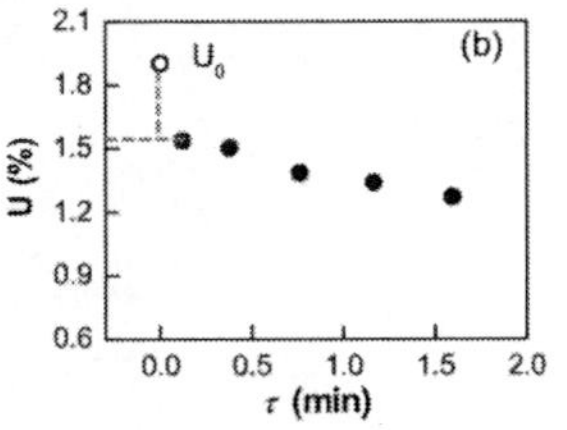

- Br_2 could also be consumed for addition reaction
- The additional rate: $r_{HBr} < r_{Br2}$

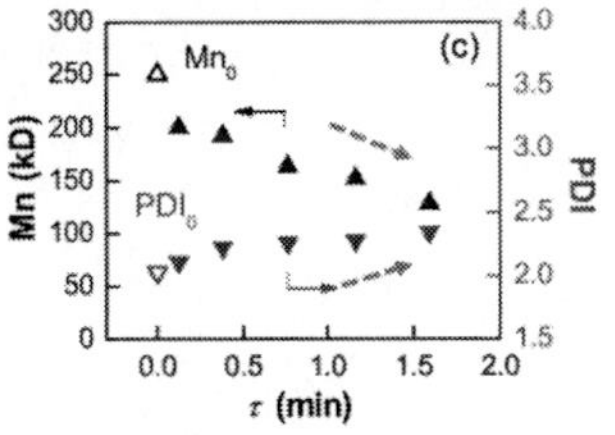

- HBr also plays a negative role in breaking the polymer chain

Results and discussion

☐ Importance of mixing intensification

x_{IIR} (wt%)	x_{Br2} (wt%)	x_{NaOH} (wt%)	$x_{Na2S2O5}$ (wt%)
10.4	6.2	0.5	0.1
Q_{IIR} (mL/min)	Q_{Br2} (mL/min)	τ (min)	T (°C)
23.8~100.2	2.4~9.9	1.44	25

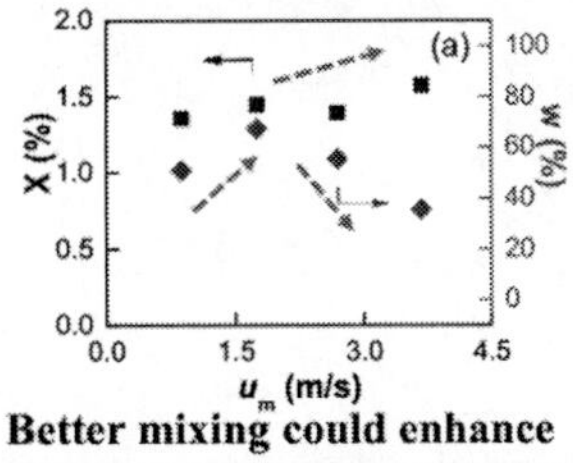

- Better mixing could enhance the conversion of substitution reaction
- $U_m < 1.6$ m/s, HBr cannot be extracted quickly by large droplets
- $U_m > 1.6$ m/s, quicker generation of HBr

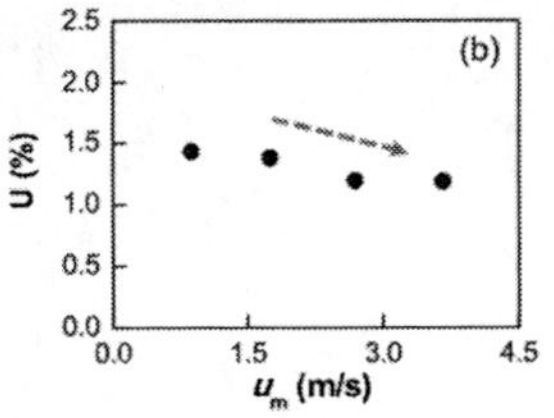

- The enhancement of mixing also intensified Br_2 addition

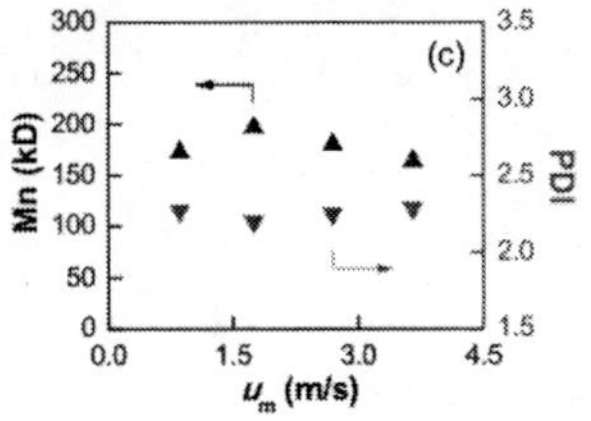

- The polymer chains break seriously in this experiment

Results and discussion

□ Changing molar ratio of reactants

x_{BR} (wt%)	x_{Br2} (wt%)	x_{NaOH} (wt%)	$x_{Na2S2O5}$ (wt%)
10.4	4.8~6.3	0.36~0.60	0.07~0.12
Q_{BR} (mL/min)	Q_{Br2} (mL/min)	τ (min)	T (°C)
85.3	8.1~10.1	1.22~1.25	25

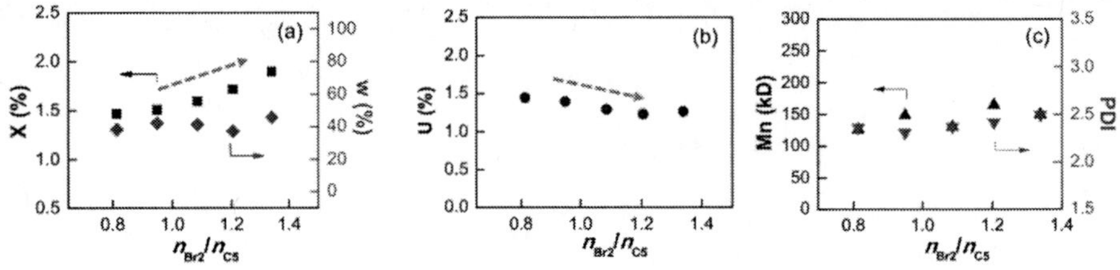

- **More Br_2 contributes to high mass concentration of substituted Br**
- **The excessive Br_2 also intensify the addition reaction**
- **High concentration of HBr keeps w and Mn at a low level**

Results and discussion

□ Effect of reaction temperature

x_{BR} (wt%)	x_{Br2} (wt%)	x_{NaOH} (wt%)	$x_{Na2S2O5}$ (wt%)
10.5	6.0	0.5	0.1
Q_{BR} (mL/min)	Q_{Br2} (mL/min)	τ (min)	T (°C)
95.3	9.7	1.51	25

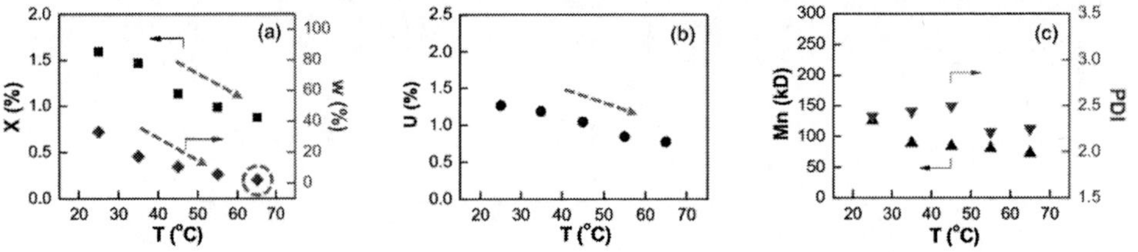

- **The high temperature accelerated the Br substitution position transfer reaction and the addition reaction**
- **The polymer decomposition was also easier at high temperature**

Results and discussion

☐ Water assistant method

x_{IIR} (wt%)	x_{Br2} (wt%)	x_{NaOH} (wt%)	$x_{Na2S2O5}$ (wt%)
9.9	6.3	0.5	0.1
Q_{IIR} (mL/min)	Q_{Br2} (mL/min)	x_{H2O} (wt%)	T (°C)
70~99.9	7.1~10.1	1.0	25

Pre-dispersed water droplets

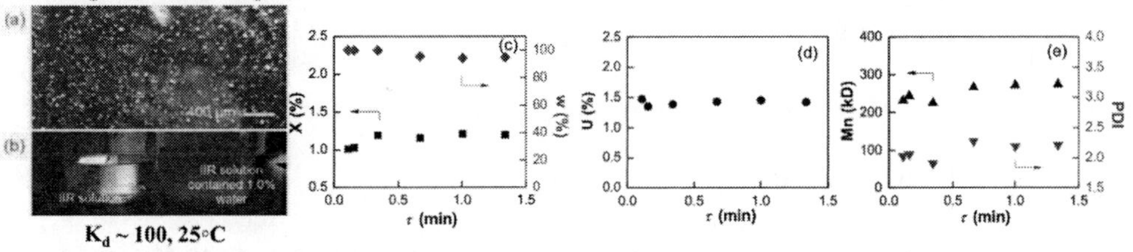

K_d ~ 100, 25°C

- The molar ratios of BIIR-1 are close to 100% even at high residence time
- More Br_2 would be extracted at long residence time with more HBr in the water phase
- This water assistant method works well for preventing the addition reaction

Outline

- **Background**
- **Experimental section**
- **Results and discussion**
- **Conclusions**

Conclusions

- **The microreactor system, which had significant advantages in reactant mixing, less than a minute residence time control and low-cost corrosion prevention, worked well for synthesizing BIIR.**
- **Low reaction temperature was found a good way to prohibit addition reaction.**
- **A water assistant technology was proposed to prevent the Br transfer reaction and the polymer chain decomposition by extracting HBr in-time from the reacting mixture.**

Thank you for your listening!

Any question will be highly appreciated!

500549 Modeling and Optimization of a Two-Dimensional Tubular Reactor for an Exothermic Reaction

Abstract:

As the world searches more energy resources at more remote locations, industry has been trying to optimize topside system of the offshore plant. One of the important aspects for this is an optimization of reactor in floating production storage and offloading (FPSO) system, where space and weight savings are crucial. In this study, a tubular reactor equipped with a cooling jacket for the exothermic reaction in topside platform was modeled based on the reaction mechanisms and kinetics obtained from experimental results. The influence of various design and operating variables (e.g. reactor length, gas flow rate, inlet temperature and concentration, and thermal conductivity) on conversion and temperature profiles along the axial and radial directions inside a reactor were studied. For this, the modeling was described in an axisymmetric coordinate system with finite element approximations using computer simulations. Lastly, the mathematical modeling of the reactor design was optimized to maximize an overall yield of the main products.

Acknowledgement
This work was supported by a research project of â€œDevelopment of Integrated Interactive Model for Subsea and Topside System to Evaluate the Process Design of Offshore Platformâ€, funded by Ministry of Trade (project No. 10060099).

490572 Advanced Control Strategy Approaches for Continuous Pharmaceutical Manufacturing: The Regulatory Landscape

Abstract:

The FDA believes that strategic efforts to encourage and sustain improvements in pharmaceutical manufacturing quality are critical to the FDA mission of making safe, effective and high-quality drug products consistently available to the patients who need them. Over the past decade since the publication of the PAT guidance, there has been significant progress in modernizing the pharmaceutical manufacturing sector. There has been a general move away from fixed uni-variate parameter specifications towards a multivariate design space approach and the adoption of in-process measurement systems to aid process development. At the same time industry and regulators are being confronted with and increasing complex challenges including unprecedented drug recalls and shortages. The implementation of advanced control strategies can help address these challenges by increasing the efficiency, flexibility, and robustness of manufacturing processes. In FDAâ€™s experience to date, continuous pharmaceutical manufacturing is serving as a driving force for modernizing control strategy approaches including model based control, multivariate monitoring, analysis of large of data sets, and Real Time Release Testing (RTRT). Process models that can capture the dynamic relationships between process parameters, raw materials, and product attributes in a quantitative and predictive manner support the development and implementation of these advanced control strategy approaches. While there are no regulatory barriers for adopting advanced control strategy approaches, there still may be hurdles encountered when fitting these approaches into the existing regulatory framework. In this presentation, scientific and regulatory considerations for advanced control approaches designed to mitigate identified risk areas are discussed including performance criteria and lifecycle considerations. Continuous drug product and drug substance manufacturing case studies are utilized to highlight these considerations.

495150 Pharmacy-on-Demand: A Continuous Revolution in Pharmaceutical Manufacturing?

Abstract:

Unlike its counterparts in the petrochemical, food and drink industries; pharmaceutical manufacturing still applies a non-continuous or â€œbatch-wiseâ€ approach. Thus, current pharmaceutical processes are ill-equipped to react to market forces or surges in demand due to humanitarian outbreaks. Active pharmaceutical ingredients (API) are world travelers with complex supply chains and long batch syntheses, meaning it can take months to manufacture and deliver the final drug product to the patient. This enormous space/time demand can result in reduced profits and inflated capital and operating expenditure.

With the era of blockbuster drugs apparently over and Pharmaceutical companies racing to the bottom of the pricing market in recent times, a paradigm shift towards continuous manufacturing may provide the answer. The benefits of flow chemistry in the synthesis of API has now been successfully demonstrated on a number of occasions. Likewise, many Drug Product unit operations are currently conducted in a continuous manner. The present challenge is now the implementation of an end-to-end manufacturing strategy that can allow these methodologies to work in concert, in a streamlined manner that enables drug substance and drug product production in one manufacturing platform.

Pharmacy on Demand (POD) seeks to address these manufacturing inefficiencies through the use of a compact and reconfigurable platform, capable of multi-step syntheses for a catalogue of API. Substantial advances in complex reaction telescoping, real time formulation, reaction engineering and pumping technologies have enabled the design, construction and implementation of the platform. Tight control of process inputs and the implementation of in-line process analytical technologies enables high volumes of APIs (Atropine, Benadryl, Ciprofloxacin, Diazepam, Doxycycline, Lidocaine and Ibuprofen) to be manufactured over a three-week campaign period. This timeframe involved the synthesis, recrystallization and formulation of each API to their correct dosage form. These solid or liquid formulated Drug Products are produced to a standard that passes all USP specifications.

496111 Platforms for Integrated Continuous Drug Subtance Design and Manufacture

Abstract:

As pressure grows on pharmaceutical companies to deliver increasing levels of process performance, reduced capital and manufacturing costs the industry is looking more and more to continuous processing to help tackle some or all of these challenges.

Within this presentation case studies will be presented where unique continuous processing hybrid platforms have been utilised for the development and operation of continuous processes. These processes included heterogeneous and homogenous chemistries, work-up/distillations, crystallization and isolation. Specifically within the case studies presented details will be outlined, via PAT integration, engineering kinetic and statistical models, experimental and lab rig design via 3D printing, required to enable the design, optimisation and operation of these platforms . During the presentation the design considerations over CSTRs, PFRs, hybrid or novel 3D printed reactor design configurations will be presented and how such design selections can impact or tailor the design properties of the intermediates and importantly the final API solid state characteristics.

496380 Residence Time Distribution and Material Traceability for Continuous Drug Substance Processes

Abstract:

Title: Residence Time Distribution and Material Traceability for Continuous Drug Substance Processes

Authors: Christopher Polster[1], Venkata Ramana Reddy[3], Carla Luciani[1], Stephen Jeffery[2], Martin Johnson[1], Kevin Chinn[4], Hod Finkelstein[4]

[1] *Eli Lilly & Co., Indianapolis, IN-46221, USA*

[2] *Eli Lilly S.A. Irish Branch, Dunderrow, Kinsale, Co. Cork, Ireland, P17 NY71*

[3] *Lilly Capability Center India, Bangalore, India*

[4] *Trutag Technologies, Inc., 2045 Lauwiliwili St, Unit 301, Kapolei, HI 96744*

Abstract

The concept of residence time distribution (RTD) to analyze chemical reactor performances appeared earlier 1930â€™s.[1] Danckwerts materialized the field by defining the most common RTDs.[2] Currently, the use of RTDs to understand mixing characteristics of the units/trains as well as material traceability and batch definition is an established approach.

With the rapidly growing adoption of continuous manufacturing processes in the pharmaceutical industry, the understanding of RTD through modeling and characterization is imperative to support both R&D activities and long term commercial manufacturing of both drug substance and drug product.[3],[4],[5]

In this work, some of the techniques used to measure RTD are described. Also, libraries were developed to provide a tool for R&D and manufacturing scientists to develop and simulate plant flowsheets that can be used for material traceability, determining diversion boundaries, and guide decision trees for diversion scenarios. Specific case studies will be shared.

[1] MacMillin, R.B., Weber, M. The Theory of Short-Circuiting in Continuous-Flow Mixing Vessels in Series and the Kinetics of Chemical Reactions in such Systems. *Transactions of American Institute of Chemical Engineers* **1935**, 31, 409-458.

[2] Danckwerts, P.V. Continuous Flow Systems. Distribution of Residence Times. Chem Eng. Sci. 1953, 2, 1-13.

[3] Manley, L. Manufacturing Platformâ€™s Residence Time Distribution Using Major and Minor Unit Formula Excipient Step Changes and Soft Sensors. *2016 AIChE Annual Meeting* (342c).

[4] Garcia-Muñoz, S., Luciani, C., Shi, Z., Roth, W., Butterbaugh, A. Hanson, J., Manley, L., Barnes, L., Leavesley, I., Almaya, A. Model-Aided Development of a Continuous Drug Product Manufacturing Process.

[5] Engisch, W., Muzzio, F. Using Residence Time Distribution (RTDs) to Address the Traceability of Raw Materials in Continuous Pharmaceutical Manufacturing. *J. Pharm Innovation* **2016**, 11, 64-81.

494238 A Miniature CSTR Platform for Continuous Processing of Multiphase Systems

Abstract:

Continuous manufacturing has emerged as a powerful tool for producing fine chemicals (e.g. active pharmaceutical ingredients (APIs)) over the past decade. However, numerous important reactions in pharmaceutical industry involve multiphase interactions, which create challenges for flow chemistry to gain a wider acceptance. For example, the presence of stoichiometric amount of solids as reagents, by-products, or products leads to the issue of blocking a flow reactor, and the problem is more severe for the milli-scale or micro-scale flow systems. Besides, the mass transfer resistance at the phase boundaries sometimes becomes the rate-limiting step. Considering current widely used tubular reactors, only providing limited interfacial interactions imposed by the laminar flow pattern, a more universal and versatile reactor design is desired for handling solids in flow, generating well-controlled mixing performance, and at the same time, offer compatibility to withstand harsh chemical environment.

Here, we present a new miniaturized continuously stirred-tank reactor (CSTR) for continuous processing of multiphase systems. The design, with the feature of plug-and-play, takes efficient mass and heat transfer, chemical compatibility, and ease to fabricate and clean into account. Single-phase residence time distribution (RTD) measurements of the CSTR unit reveal nearly ideal CSTR mixing. It also has the capability to tune the RTD by connecting several units in series to achieve approximate plug flow reactor (PFR) RTD, which gives the flexibility to optimize the selectivity based on the reaction kinetics.

Two versions of miniature CSTRs were developed for: (i) solid handling in flow, and (ii) high-performance liquid-liquid mixing.

Solid handling CSTR: employing the non-stick polytetrafluoroethylene (PTFE) as the reactor body material with minimized inter-unit connection enables the CSTR to continuously handle high-concentrated slurries for hours without significant signs of clogging. Two solid-forming reactions, (i) glyoxal reacting with cyclohexylamine to form N,Nâ€²-dicyclohexylethylenediimine, and (ii) sulfonylation of 2-octanol with methanesulfonyl chloride, demonstrate its practical usefulness as a powerful for flow chemistry field.

High-performance liquid-liquid CSTR: incorporating the magnetic-coupling as the driving force to provide external agitation inside the confined reactor space gives the possibility to generate extraordinary two-phase mixing performance without losing the capability to seal the reactor. We demonstrate the reactor with a comparable liquid-liquid mass transfer coefficient to Corning Advance-flow Reactor, which is the state-of-the-art among commercial reactors. More importantly, it gets around the problem of coupling of flowrate and mixing, and is capable to accommodate reaction kinetics from seconds to hours. Phase transfer catalyzed reactions were demonstrated in this newly designed CSTR, with the corresponding kinetic model developed for Makosza phase-transfer mechanism. In addition, the success of carrying out biphasic Suzuki-Miyaura reactions further validates its suitability for the pharmaceutical applications.

Grams of API to Commercial-Scale Continuous GMP Drug Product Manufacture in Less Than a Year

Carl Hartmann

Lilly

Phase I Clinical Supply Options

Phase I Clinical Demand: initial forecast **200,000+** doses

Drug-in-capsule
Output: 200 capsules / hr
Consider: RBA study, low throughput

Small-Scale Development
Req: 250 g API

Continuous Tablet Dev
Req: under 2 kg API
Output: 50k-100k /hr

CT Manufacture Decision

?

Time

2

Small-Scale Development Defined

- Dispense, Blend, Clean

Lubrication Scale Model

3+ hrs and 200-10,000 g API → 30 min and 1-20 g API

- Setup, Compress, Clean

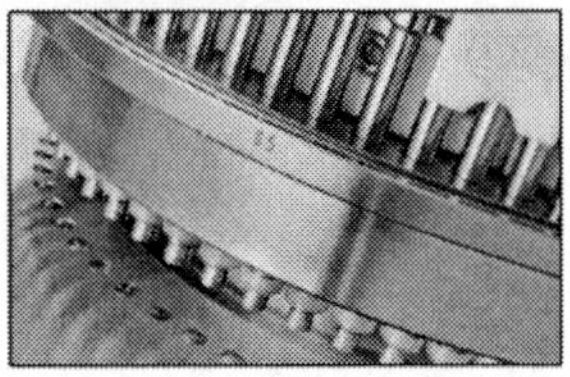

Intrinsic Material Properties Models

3+ days → 1 hr

3

Small-Scale, but Big Data

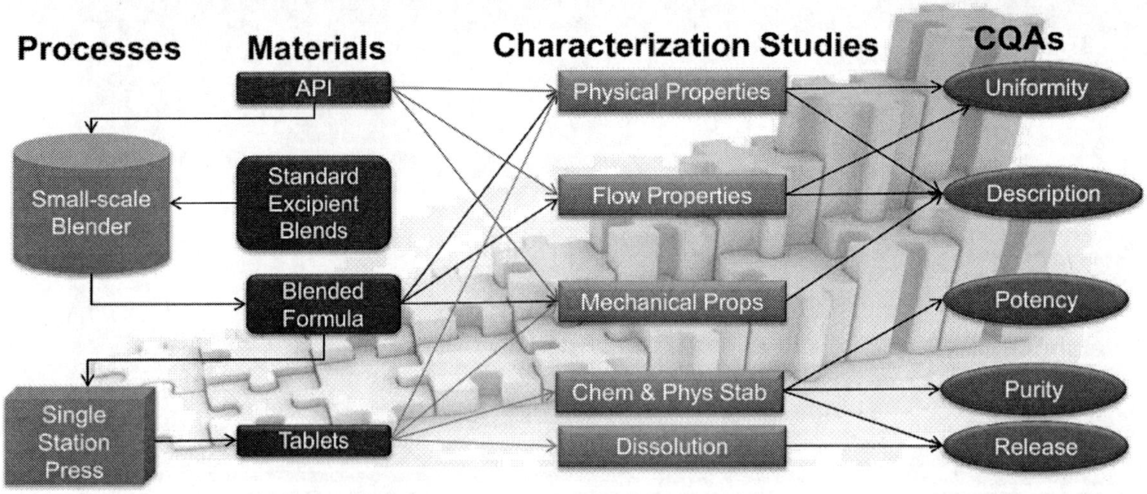

4

Small-Scale Development Blending

DS Lot	X90 (um)	Hausner Ratio	FFC (flow) values
A	192.4	1.41	4.10
B	179.8	1.46	9.18
C	129.7	1.41	4.86
D	135.7	1.34	4.84
E	72.2	1.56	2.85

Five drug substance (DS) lots of varying particle size distributions (PSD), each 50 gram quantity

Each DS lot blended into a common formulation for comparison, each blend targeting 50 gram quantity

Material Name	Common Formulation (%)
Drug Substance	28
Lactose Monohydrate	16
Microcrystalline Cellulose	48.5
Superdisintegrant	5
Lubricant	2.5

5

Small-Scale Development Tablets

- Standardized tablets:
  - 6 mm round, flat, 100 mg target weight

A preliminary blend at 42% drug load (DL). Observation of picking/sticking.

Reduced drug load

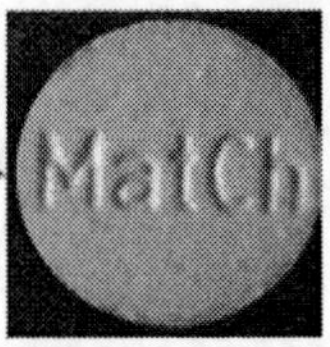

Comparison of five differing PSD DS lots conducted in blends at 28% DL. No visual picking/sticking.

- Single station instrumented press
- Compression profile and stability testing

6

Initial Tablet Results - Common Blend

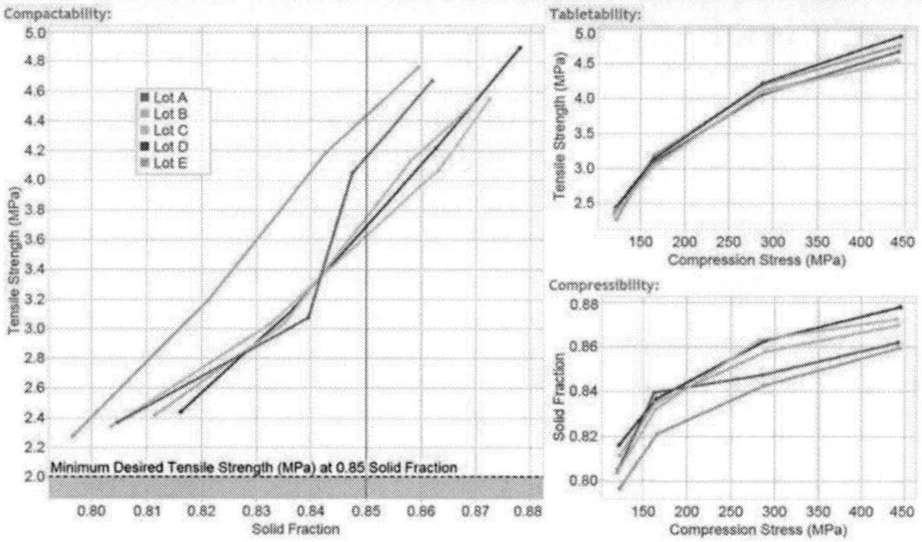

7

Adjustments and confirmation

- Initial risk assessment:
  - Smallest API, poor flow
  - No other red flags
- Formulation development:
  - Reduce drug load
  - Eliminate lactose

8

CM Direct Compression Flow Diagram

FOYA 2017
Facility of the Year Awards
CATEGORY WINNER
Process Innovation

FOYA 2017
Facility of the Year Awards
CATEGORY WINNER
Facility of the Future

FOYA 2017
Facility of the Year Awards
OVERALL WINNER

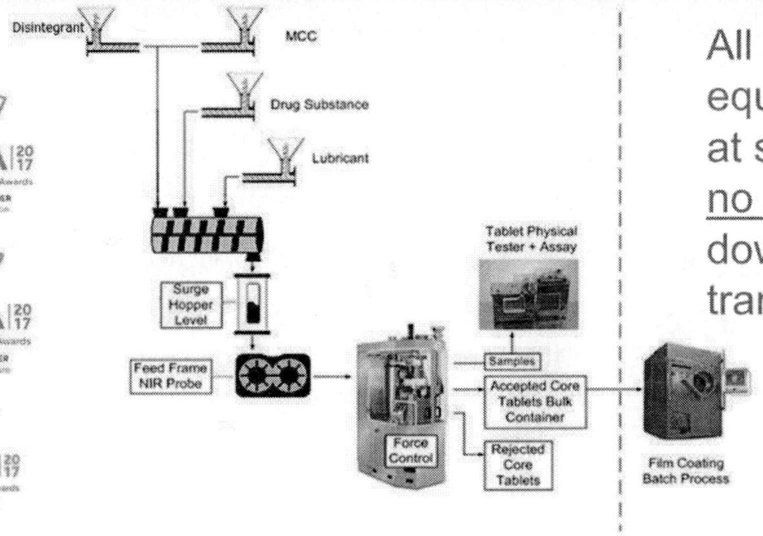

All CM equipment trains at same scale; <u>no scaling</u> up or down with tech transfer

9

CM Direct Compression Animation

10

Leading into Demonstration Batches

- Off-line **feedability** (LIW feeder) for DS:
  - For twin-screw tooling selection
- Spectroscopy: **NIR** signal range of DS and DS blends
  - For tablet press feed frame NIR on-line monitoring
- Select **tablet tooling** shapes and sizes appropriate for CT
- Identify **coating** system and color appropriate for CT

5 mg strength
6 mm round
Pink coat

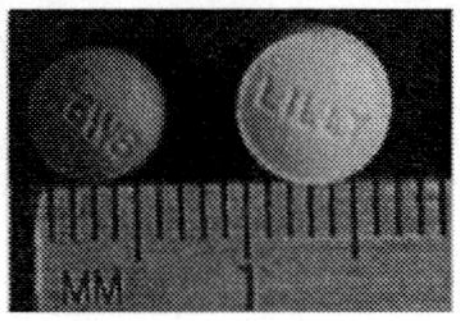

25 mg strength
7 mm round
Yellow coat

11

Development Demonstration Batches

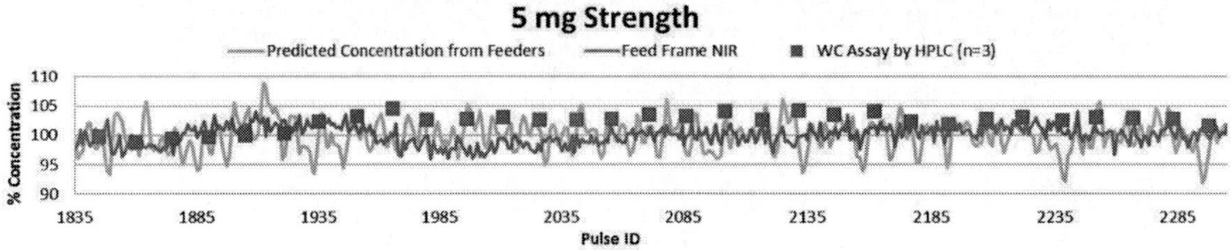

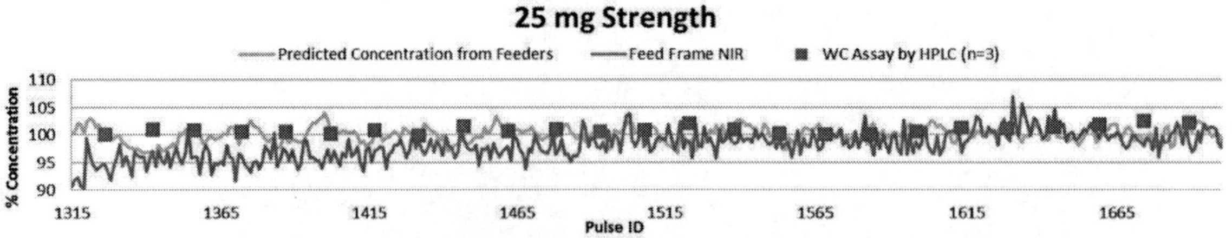

Less than 2 kg API required in total for both batches

12

Phase I Clinical Supply Decision

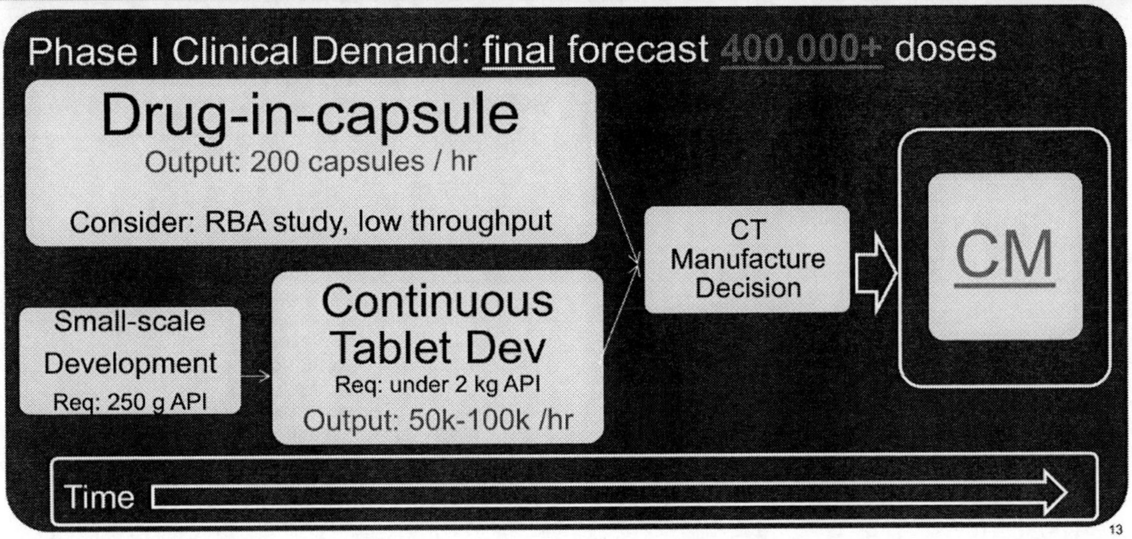

13

GMP Commercial Site Manufacture

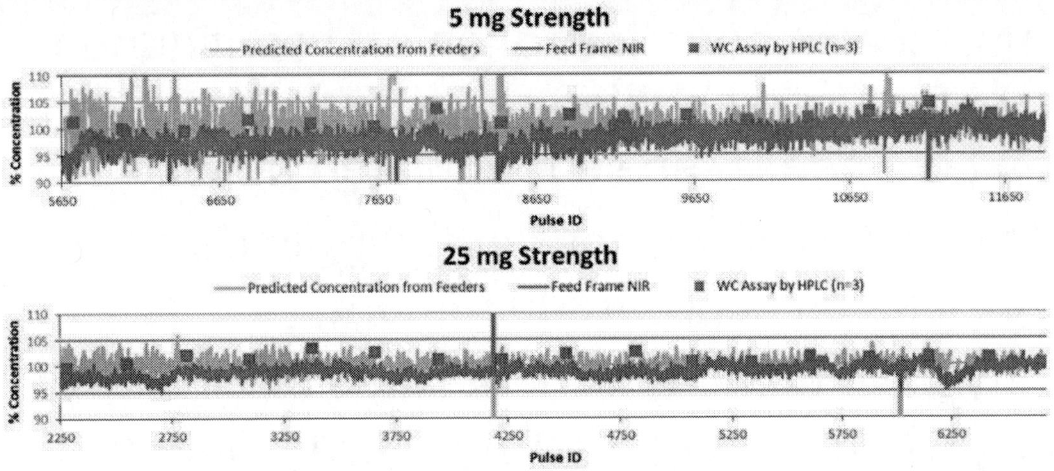

Both batches readily meet USP <905> and ASTM 2810 criteria

14

Timeline

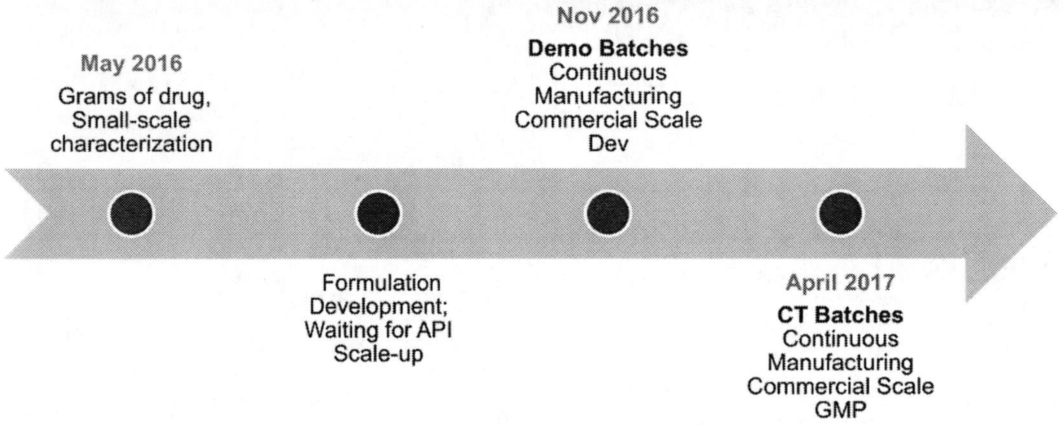

Grams to GMP commercial-scale in less than a year!

15

Conclusions

- Small-scale characterization and continuous manufacturing enables **rapid development** of a commercial-scale formulated tablet
- CM platform allows for **flexibility** in delayed decisions on supply requirement
- Early development work assures feasibility of CM platform to **buy down risk** with little API

16

Acknowledgements

Mark Polizzi
Dave Myers
Mike Miller
Pete Shi
Adam Butterbaugh
Josh Hanson
Melissa Keeney
Joe Martinelli
Adam McFarland
Moussa Boukerche
Dale Greenwood
Kirk VanderKamp
Tim Kramer
Sarah Schmitt

17

Questions?

18

487680 A Continuous and Controlled Pharmaceutical Freeze-Drying Technology for Unit Doses

Abstract:

Driven by growing needs in the biopharmaceutical market and regulatory pressure, a continuous and controlled freeze-drying technology for unit doses to preserve biopharmaceuticals has been developed. Such continuous process allows a more efficient, cheaper, greener and controllable manufacturing method compared to traditional batch production systems, offering competitive advantages and business opportunities.

Pharmaceutical freeze-drying (lyophilization) is a low-temperature drying process in which aqueous solutions of heat-labile biopharmaceuticals are converted into solids with sufficient stability for distribution and storage. Similar to all manufacturing processes of drug products (solids, semi-solids and liquids), conventional pharmaceutical freeze-drying is generally accomplished using batch processing that is considered time-consuming, costly, non-flexible and lacking robust quality control and real-time release.

Four major industrial drivers are demanding a more efficient and better controllable pharmaceutical freeze-drying technology for unit doses: cost-cutting, regulatory pressure, a fast growing biopharmaceutical market and an ageing population requiring more personalized medicines.

The continuous and controlled freeze-drying technology, developed following the principle of model based design, offers clear advantages over current batch production such as cost reduction (up to 50%), track-and-trace product quality control, and a significant reduction of processing time (> 40 times faster, e.g. 1 hour instead of 5 days at a vial level), reduced need for clean room and a substantial sustainability gain.

493330 Experimental and Simulation Studies of High-Temperature Ethane Dehydrogenation in Microporous Zeolite Membrane Reactor

Abstract:

Experimental and Simulation Studies of High-Temperature Ethane Dehydrogenation in Microporous Zeolite Membrane Reactor

*Shailesh Dangwal, Ruochen Liu, Seok-Jhin Kim**

School of Chemical Engineering, Oklahoma State University,

420 Engineering North, Stillwater, OK 74078, USA

ABSTRACT

Ethylene is an important chemical engineering ingredient and used in polymerization, oxidation, and alkylation. The increasing demand for ethylene has stimulated substantial research into the development of new processes to reduce energy consumption. Catalytic ethane dehydrogenation using a membrane reactor is an attractive solution because the cracking equilibrium can be shifted in favor of ethylene by selective removal of hydrogen. We reported the intensification of ethane dehydrogenation reaction in MFI zeolite membrane reactors operating with a Pt/Al_2O_3 catalyst. The effects of the reaction conditions and the membrane separation properties on the ethane conversion in the membrane reactor were investigated experimentally and by simulations using a simple one-dimensional model which was validated by the experimental data. It was demonstrated that the rate of H_2 membrane permeation significantly enhances the ethane conversion. MFI membrane reactors allow the equilibrium limit of ethane conversion to be surpassed at high temperatures. The model calculations have shown that near-completion ethane conversion (>99.5%) may be achieved under realistic operating temperature, pressure, space velocity and catalyst load even for membranes with moderate H_2 selectivity, ($aH_2/C_2H_4<10$) and H_2 permeance ($<1.8\times10^{-7}$ mol.s.m^2.Pa).

*Corresponding author: E-mail: seokjhin.kim@okstate.edu

501253 Experimental and Numerical Study of an Intensified Water-Gas Shift (WGS) Reaction Process Using a Membrane Reactor (MR)/Adsorptive Reactor (AR) Sequence

Abstract:

Reactive separation processes, such as membrane reactors (MRs) and adsorptive reactors (ARs), have been attracting increased interest for industrial applications. Such a hybrid system combining a membrane reactor (MR) and an adsorptive reactor (AR) in tandem (with the AR following the MR, and the MRâ€™s reject stream serving as the ARâ€™s feed) can produce an ultra-pure H_2 product (without the need for using a post-processing step), and can deliver WGS reaction conversion exceeding equlibrium. The proposal configuration (AR following MR) provides significant flexibility in addition to high-purity hydrogen and carbon dioxide production.

The primary objective of this work is to first develop and simulate an advanced, and detailed mathematical model for the MR/AR sequence and then validate this model using experimental data obtained at the bench-scale. The MR system is composed of a reaction zone packed with catalyst and quartz pellets, and a water/hydrogen permeation zone. In the AR, both catalyst and adsorbent materials are used for simultaneous reaction and separation. The AR is a dynamically operated process. Both steady-state and dynamic experiment are carried out in a laboratory setting to capture the individual MR/AR performances. Experiments are carried out for a broad range of temperatures (up to 300 ^{o}C) pressures (up to 25 bar) relevant to realistic application conditions. The velocity, temperature and species concentration profiles in the MR reaction and permeation zones are captured by momentum/species transport models accounting for convection/reaction /diffusion/conduction mechanisms. The rigorous Maxwell-Stefan and dusty gas models are applied to describe mass diffusion fluxes. Finally, the performances of the combined system (the membrane reactor (MR) followed by an adsorptive reactor (AR) is simulated for a broad range of the operating conditions and design parameters. The simulation results are compared with the experimental findings.

[HC1]and pressures
[HC2]experimental

493066 Synthesis & Performance Study of Pd-Au/PSS Membrane Reactor to Produce Highly Pure Hydrogen Via Natural Gas Steam Reforming

Abstract:

One possible use of hydrogen, without direct greenhouse gas emissions, is as feed for a fuel cell (FC), with the most readily available technology being a proton exchange membrane FC (PEMFC). In order to avoid poisoning of PEMFCâ€™s Pt-based catalyst due to the presence of ppm levels of CO, the hydrogen feed needs to be ultra-pure. The industrial process for hydrogen production, which is a multi-step energy-intensive process followed by further separation/purification, can be a potential source [1]. However, as an alternative method, a Pd-based membrane reactor (MR) may be used due to its ability to provide high-purity hydrogen directly. Moreover, the MR works at milder operating conditions compared to a traditional system.

In recent years, Pd-based composite membranes, i.e., thin metallic layers supported on porous materials such as ceramics or stainless steel, have been considered due to their lower cost (thin Pd layer) and higher mechanical resistant (porous support) compared to dense Pd-based membranes [2,3]. Moreover, the use of Pd-alloys, in particular, Ag, Au, Pt, has grown in the recent years given their ability to resist sulfur poisoning in addition to their enhanced hydrogen permeability [4]. Therefore, the aim of this study was to fabricate a Pd-Au membranes supported on porous stainless steel (PSS) with the intent to produce high-purity hydrogen from natural gas steam reforming. The membrane used in this study was fabricated via an electroless plating technique for depositing Pd and electro-plating for the deposition of Au.

The initial characterization of the membrane by way of ideal selectivity took place at 400 °C with H_2, He and N_2, and with DP in the range of 1.5 - 3.0 bar. After ideal selectivity characterization of the Pd-Au/PSS membrane, the natural gas steam reforming reaction was carried out and the MR performance in terms of CH_4 conversion, hydrogen production and hydrogen permeate purity has been analyzed and will be discussed.

References

[1] Rostrup-Nielsen, J.R., Catalytic steam reforming. 1984: Springer.

[2] S. Liguori, A. Iulianelli, F. Dalena, P. Pinacci, F. Drago , M. Broglia, Y. Huang, A Basile, "Performance and long-term stability of Pd/PSS and Pd/Al_2O_3 membranes for hydrogen separation", *Membranes*, 4 (2014) 143-162.

[3] B. Anzelmo, J. Wilcox, S. Liguori, "Natural gas steam reforming reaction at low temperature and pressure conditions for hydrogen production via Pd/PSS membrane reactor", *Journal of Membrane Science*, 522 (2017) 343-350.

[4] Chen, C.-H. and Y.H. Ma, "The effect of H_2S on the performance of Pd and Pd/Au composite membrane". *Journal of Membrane Science*, 362 (2010) 535-544.

497163 Non-Isothermal CFD Study of Ethanol Steam Reforming in a Catalytic Membrane Reactor

Abstract:

Non-isothermal CFD Study of Ethanol Steam Reforming in a Catalytic Membrane Reactor

Rui Ma[1], Bernardo Castro-Dominguez[2], Anthony G. Dixon[1], Yi Hua Ma[1]

[1.]Center for Inorganic Membrane Studies, Department of Chemical Engineering, Worcester Polytechnic Institute, 100 Institute Road, Worcester, MA 01609, USA.

[2.] University of Limerick, Sreelane, Castletroy, Co. Limerick, Ireland

Single unit hydrogen production and separation achieved by conducting ethanol steam reforming (ESR) in a catalytic membrane reactor (CMR) has attracted more interest due to the carbon neutral property of bio-ethanol, and the high reactant conversion due to process intensification.[1] In ESR process, temperature control helps to achieve higher conversions. Compared with traditional packed bed reactors (PBR), heat supply for CMR needs to compensate for the enthalpy that hydrogen carried across the membrane during the permeation process, in addition to the heat loss from endothermic reactions. Furthermore, hydrogen flux through the membrane is temperature dependent as the hydrogen permeability follows Arrhenius law. In this work, a comprehensive heat and mass transfer study of an ESR process in a membrane reactor is carried out using an experimentally-validated non-isothermal CFD model. Retentate and permeate domains were modeled by introducing a negative and positive hydrogen flux term respectively, using Sieverts' law. In order to increase the hydrogen permeation driving force, a purge gas was applied to the permeate side. Effective thermal conductivity and diffusivity were introduced to the simulation in order to take into account the catalyst bed.

A "cold spot" adjacent to the membrane was observed (Fig 1) since this highly endothermic process benefits from hydrogen removal adjacent to the membrane. From the reaction rate study of the ESR process, the process takes place very rapidly, and a slight reverse reaction for methane steam reforming was observed due to the decrease in equilibrium conversion at lower temperature. This phenomenon is most noticeable within the "cold spot". A clear hydrogen generation and depletion along the reactor on the retentate side was observed. In addition, the reactions reach equilibrium shortly after reactants enter the reactor which leads to an inefficient use of the reactor and the membrane surface area; thus, counter-current flow of the purge gas is studied, which helps with shifting the reaction equilibrium throughout the entire reactor. Using this model, the effect of temperature and pressure on hydrogen recovery and yield are studied in both co-current and counter-current cases, and a process optimization study was carried out.

Fig 1. Temperature distribution in the reactor at LHSV=3.77h-1, Pret =6 bar, S/E=5 and Tin=723K on both permeate and retentate side.

1) Ma, R., Castro-Dominguez, B., Mardilovich, I.P., Dixon, A.G. and Ma, Y.H., "Experimental and Simulation Studies of the Production of Renewable Hydrogen through Ethanol Steam Reforming in a Pilot-Scale Catalytic Membrane Reactor," Chem. Eng. J., 303, 302-313 (2016).

511513 Heat Integrated Steam Reforming Membrane Reactor

Abstract:

A heat integrated steam reforming membrane reactor was investigated for development of a gasoline fuel processor for fuel cell vehicles. This reactor utilizes Le Chatelierâ€™s Principle for driving thermodynamically equilibrium limited steam reforming reactions at higher pressures. Steam reforming at higher pressures also reduces the palladium alloy membrane area required. Coupling steam reforming with a hydrogen selective membrane provides essentially pure hydrogen for optimal fuel cell performance.

Engineering models were used extensively to help define and refine the fuel processor design. The catalytic heat exchange reactor was coupled to a separate palladium-alloy membrane and used recycle of hydrogen depleted reformate to achieve high efficiencies at moderate operating temperatures and pressures (650Â°C, 20 Bar). Recycle was achieved by using a steam ejector. Catalytic combustion of the hydrogen depleted reformate with lean air fuel cell cathode exhaust (11% O2) supplied the heat required for steam reforming. Our modeling suggested that our steam reforming approach could have thermodynamic efficiencies of up to 40%.

Screening studies were undertaken to identify high activity noble metal steam reforming catalysts. A bimetallic Ru-Ir catalyst was found to have exceptionally high activity and stability. Order-of-magnitude, improvements in catalyst performance and reactor geometry were made.

In laboratory studies excellent heat transfer was observed when conducting catalytic combustion and steam reforming co-currently on opposing walls of the wash coated reactor tubes. Several membrane geometries were used. Hydrogen permeate was obtained from a variety of feeds, including isooctane, naphtha and methane, at reaction temperatures of 600-700°C. The experimental results were in close agreement with the modeling predictions.

While considerable research and development would be required to advance this technology, these results demonstrate the potential of the heat integrated steam reforming membrane reactor system.

486015 Improving Dehydrogenation Conversion and Selectivity By Utilizing Thermally Stable Membranes

Abstract:

In an equilibrium-limited alkane dehydrogenation reaction, due to its endothermic nature, high temperature is demanded to reach high conversion. Removal of product H_2in a dehydrogenation reaction shifts the equilibrium forward to realize a higher conversion at mild conditions. Thermodynamic computation demonstrates that with the assistance of a membrane separation unit, equilibrium limit can be exceeded in alkane dehydrogenation by selective removal of hydrogen. By operating at a lower temperature, high catalyst performance can be achieved; while catalyst coking is less likely to occur. In this research, dehydrogenation of methylcyclohexane to toluene is studied as a model reaction.

To study the impact of membrane utilization on the dehydrogenation process, blend polymer materials with high thermal stability, chemical resistance and transport properties has been selected as membrane materials. The permeabilities of dense films of Polybenzimidazole (PBI) and Matrimid® blends were evaluated from room temperature to 300 C. Permeabilities increased exponentially with temperature and the selectivities decreased. All blend materials were stable over the multi-day testing.

The potential impact of selective hydrogen removal via thermally-stable polymer blends was calculated from finite element modelling. The model incorporated the measured permeation properties, measured reaction equilibrium information, and catalyst performance (kinetics and coking). As the process can be operated at a range of temperatures, the influence of system temperature on conversion, permeation rate, and catalyst deactivation is analysed. The use of a single membrane unit inter-stage between two packed beds has the potential to increase reactant conversion by 35% compared to operation without inter-stage conversion.

494458 Integration of Synthesized Catalysts and Membrane in Packed and Fluidized Bed Membrane Reformer for Production of Hydrogen through Steam Reforming

Abstract:

Hydrogen is the only energy carrier which can cater to the rising power demands across the world without compromising with the environmental impact. Its high efficiency to generate power comes in conjunction with fuel cells that can produce a desired output with absolutely no carbon emissions. However, the hydrogen fed to fuel cells must be of high purity to ensure stable performance of Pt-based catalysts of fuel cell anode. Therefore, production of hydrogen with high purity is considered crucial. In the current work, hydrogen generation is targeted with methanol steam reforming using synthesized Cu-based catalysts. To achieve high hydrogen purity, Pd-based dense supported membranes were employed after their synthesis, characterization and testing. The target of this work is to suitably integrate the catalysts with the membranes in one single unit. Such integration will cause continuous permeation of hydrogen through a H_2 selective Pd membrane. Disturbed reaction equilibrium will therefore cause an increase in the forward reaction rate according to Le Chatelierâ€™s rule and hence improve methanol conversion. This system of simultaneous generation and separation of hydrogen in one single unit is termed as a membrane reformer (MR). Catalyst placing relative to the membrane determines the reaction residence time in a membrane reformer and thereby plays a key role. Further, this factor is inter-related with reaction kinetics, hydrogen mass transfer rate from reaction zone to membrane, and lastly membrane permeation flux and perm-selectivity. In order to observe the effect of these parameters on total system performance, the membrane reformer study was paralleled between a packed bed and a fluidized bed MR. Packed bed consists of a catalyst bed (~bed ht. 2 cm) in the shell of the reactor with membrane in its lumen separated by a distance of approximately 15 cm. In contrast to that, fluidized bed gives the advantage of catalyst being fluidized in the same shell that brings it in close contact with the membrane.

The current study would determine performance of packed bed MR with varying catalyst positions and parameters such as temperature, flow rate and weight hour space velocity (WHSV). Based on these operating conditions, fluidization regime was characterized by u/u_{mf} in the range of 1.4â€“3.6 (where u_{mf} corresponds to the minimum fluidization velocity). All testing was performed in an in-house designed test rig. Dense Pd-based membranes were prepared using modified electroless deposition approach on porous SS (PSS) with yttria stabilized zirconia (YSZ) as the intermediate layer. YSZ minimizes the inter-metallic diffusion between Pd and SS. Further, catalysts with varying bi-metallic compositions were synthesized by incipient wetness impregnation. Optimization of standalone membrane as well as catalyst performance was performed prior to the integration.

Once integrated, the packed bed membrane reformer (MR) showed >95% methanol conversion and 45% H_2 recovery at 723 K and 3 bar. Further, a comparison of packed bed MR was made with a fluidized bed MR. It was observed that with increasing u/u_{mf} greater contact between the catalyst and the membrane enhances the hydrogen yield and thereby leads to better separation in comparison to packed bed. Moreover, a complete description of these results will be elaborated in terms of effectiveness factor, reaction kinetics and hydrogen permeation flux through the membrane.

495978 Technoeconomic Evaluation of Cyclosporine Crystallisation Intensification Using a Cascade of M.S.M.P.R. Crystallisers

Abstract:

Continuous Pharmaceutical Manufacturing (CPM) has been recognised as a promising new production paradigm over the currently implemented batch methods, which imply high volumes of waste and inefficiencies [1]. Whilst the literature demonstrates many continuous flow syntheses of active pharmaceutical ingredients [2], technological innovation is required to realise the benefits of integrated end-to-end CPM processes, including purification and separation unit operations. Process systems engineering (PSE) methodologies allow the screening of candidate CPM processes for their feasibility and viability prior to committing to significant time and financial investments in experimentation and scaleup [3]. Recent efforts in modelling and optimisation of the upstream CPM of APIs have demonstrated the importance of conducting PSE studies in establishing economically viable separation processes [4], realising issues in modelling of particulate processes for solids handling in continuous operation [5] and establishing limitations attainable process improvements in pharmaceutical manufacturing [6].

Cyclosporine is an immunosuppressant drug for the prevention of organ transplant rejection. The continuous cooling crystallisation of cyclosporine has recently been investigated using continuous mixed suspension, mixed product removal (MSMPR) crystallisers, due to their broad applicability and ease of integration into existing pharmaceutical processes [7]. The effect of process configuration [8], mother liquor and solids recycle [9], number of stages [10] on crystallisation yields, purities and product size distributions have been studied to demonstrate the feasibility of continuous crystallisation of the API. Technoeconomic comparisons between batch and continuous crystallisation configurations for cyclosporine to validate the viability of continuous crystallisations are yet to be implemented.

The present paper work develops a steady-state process model for the continuous cooling crystallisation of 100 kg per annum of cyclosporine from acetone including a systematic technoeconomic comparison of various continuous process configurations with the batch method, with a focus on computationally exploring and demonstrating process intensification. The effect of the number of MSMPR stages and varying recycle considerations (mother liquor versus solids recycle) on attainable yields, process material efficiencies and costs are compared. Mother liquor recycle requires vacuum evaporation of solvent, whilst solids recycle uses gravity driven separation to allow recycle of concentrated slurry to MSMPR stages. The developed model simultaneously solves MSMPR mass balances, crystal population balances and crystallisation kinetics to calculated stage and overall crystallisation yields. Prediction of temperature-dependent cyclosporine solubilities in acetone uses a regressed surrogate polynomial established from published experimental data [11]. Temperature- and concentration-dependent growth and nucleation rates are computed on the basis of published kinetic parameters [10] for accurate calculation of attainable yields. Material efficiencies are quantified via popular green chemistry metrics, namely the process mass intensity (PMI) and the mass productivity (MP), which offer quantitative insight into process intensification potential, for each configuration considered. Unit volumes and material throughputs required to meet the plant API capacity are scaled to account for stage inefficiencies. Economic analyses are conducted via an established methodology for pharmaceutical manufacturing processes [12].

The continuous MSMPR crystallisation configurations investigated show potential for application in the CPM of cyclosporine from the calculated attainable material efficiencies and total costs. The sensitivity of process performance to various operating parameters has been explored as well, indicating the significance of rigorous design space investigation of CPM processes. Our work encourages subsequent technoeconomic studies and life cycle assessments (LCA) of the CPM of cyclosporine for further investigation of process feasibility and viability.

The present study exhibits the importance of conducting PSE studies of candidate continuous separation processes for CPM prior to further process development. We demonstrate the economic viability of CPM to facilitate the transition from batch methods, and quantify the potential intensification benefits of improved material efficiency of continuous separation processes in comparison to their batch counterparts. This study expedites further process modelling for the CPM of cyclosporine and other candidates, as well as optimisation of continuous separation processes to realise the full benefits of continuous operation in the pharmaceutical industry.

LITERATURE REFERENCES

[1] Poechlauer, P., Colberg, J., Fisher, E., Jansen, M., Johnson, M. D., Koenig, S. G., Lawler, M., Laporte, T., Manley, J., Martin, B., O'Kearney-McMullan, A., 2013. Pharmaceutical roundtable study demonstrates the value of continuous manufacturing in the design of greener processes, *Org. Process Res. Dev.* **17**(12): 1472–1478.

[2] Baumann, M., Baxendale, I. R., 2015. The synthesis of active pharmaceutical ingredients (APIs) using continuous flow chemistry, *Beilstein J. Org. Chem.* **11**: 1194–1219.

[3] Teoh, S. K., Rathi, C., Sharratt, P., 2015. Practical assessment methodology for converting fine chemicals processes from batch to continuous, *Org. Process Res. Dev.* **20**(2): 414–431.

[4] Diab, S., Gerogiorgis, D. I., 2017. Process modelling, simulation and technoeconomic evaluation of separation solvents for the Continuous Pharmaceutical Manufacturing (CPM) of diphenhydramine, *Org. Process Res. Dev.* DOI: 10.1021/acs.oprd.6b00386, in press.

[5] Rogers, A., Ierapetritou, M., 2014. Challenges and opportunities in pharmaceutical manufacturing modelling and optimization, *Comput. Aided Chem. Eng.* **34**: 144–149.

[6] Benyahia, B., Lakerveld, R., Barton, P. I., 2012. A plant-wide dynamic model of a continuous pharmaceutical process, *Ind. Eng. Chem. Res.* **51**(47): 15393–15412.

[7] Su, Q., Nagy, Z. K., Rielly, C. D., 2015. Pharmaceutical crystallisation processes from batch to continuous operation using MSMPR stages: Modelling, design, and control, *Chem. Eng. Process. Process Intensif.* **89**: 41–53.

[8] Alvarez, A. J., Singh, A., Myerson, A. S., 2011. Crystallization of cyclosporine in a multistage continuous MSMPR crystallizer, *Cryst. Growth Des.* **11**(10): 4392–4400.

[9] Li, J., Trout, B. L., Myerson, A. S., 2016. Multistage continuous mixed-suspension, mixed-product removal (MSMPR) crystallization with solids recycle, *Org. Process Res. Dev.* **20**(2): 510–516.

[10] Li, J., Lai, T. C., Trout, B. L., Myerson, A. S., 2017. Continuous crystallization of cyclosporine: the effect of operating conditions on yield and purity, *Cryst. Growth Des.* **17**(3): 1000–1007.

[11] Wong, S. Y., Tatusko, A. P., Trout, B. L., Myerson, A. S., 2012. Development of continuous crystallization processes using a single-stage mixed-suspension, mixed-product removal crystallizer with recycle, *Cryst. Growth Des.* **12**(11): 5701–5707.

[12] Schaber, S. D., Gerogiorgis, D. I., Ramachandran, R., Evans, J. M. B., Barton, P. I., Trout, B. L., 2011. Economic analysis of integrated continuous and batch pharmaceutical manufacturing: A case study, *Ind. Eng. Chem. Res.* **50**(17): 10083–10092.

489029 Application of Advanced Process Control Tools in Continuous Processing

Abstract:

Continuous manufacturing is now widely discussed in the pharmaceutical industry. The availability of turn-key lines offered by various suppliers allows the transfer of established and new products to the continuous manufacturing approach. At the moment the focus is mainly on solid oral dosage forms including the common routes via wet and dry granulation or direct compaction. From a processing point-of-view the shift to continuous mode is supported by the fact that most common unit operations are continuous machinery by design, such as feeding, tableting or granulation. One challenge is the integration of all units into a single line, which needs implementation of interfaces, transport units and buffers. The remaining challenge is the establishment of control concepts that enable a fully automated, yet still flexible process, capable of handling start-up and shut down, as well as of out-of-spec events. Tools for in-line monitoring of critical quality attributes and real-time acquisition of process data have been widely demonstrated. Process models, fast enough for design of controllers, have been developed recently. However, most continuous processes filed by pharmaceutical industry still remain on a level 2 control strategy based on in-process controls without control loops.

In this work different tools for advanced process control were developed and investigated and their implementation in processes or unit operations will be shown. This includes novel process analytical technologies enabling real-time data acquisition of statistical representative numbers or even each single dosage form, concepts for out of spec material handling based on modelling of residence time distribution or model predictive control. The latter approach was developed for a feeding and blending unit. The rotation speed of the blender is manipulated to control the mass hold-up and the outlet mass flow rate in order to mitigate fluctuations of homogeneity caused by feeder events. Another example deals with the discharge of out-of-spec material, in combination with an operating concept for a tablet press based on different pre-defined set points of the turret speed. Recent developments in the field of PAT will complete the presentation.

491286 Nonlinear First-Principle Model-Based Control of a Continuous Pharmaceutical Manufacturing Process

Abstract:

Model-based process control is one of the pillars of continuous pharmaceutical manufacturing process by the QbD approach. The first-principle approach, which comes from in-depth understanding of the governing transport phenomena and performance of unit operations, yields developing of predictive models with which the continuous processes can be manipulated, controlled, and optimized. Despite empirical modeling, where limited operation conditions and process status are being mapped within a DoE, other unseen nonlinearities in the process can be captured in a nonlinear first-principle model to optimally control the process.

In this work, a nonlinear first-principle distributed dynamic model was developed to control a drying process for continuous manufacturing of orally dissolving films loaded with a pharmaceutical compound. The drying model is the core of active control for the process, in which the process parameters (such as air temperature, or production rate), feed compositions (like solid content and concentration), or final productsâ€™ properties (for instance the required final moisture content) can be the manipulating variables. Simulation of transport phenomena, diffusion controlled mass transfer (mutual solvent-polymer vs. moisture sorption isotherm), the spatial distribution of solvent concentration in the film and the moving boundary on the interface due to the thin film shrinkage during drying are some of the complexities of the model development. The final quality attribute of the product is set as the control set point and other process variables were manipulated in the nonlinear model control (MIMO) to converge the process in the design space of variable inputs. The model funds the steps for future phases of PAT implementation, RTRt, and process optimization.

497129 A Fault-Tolerant Control Design for Real-Time Release in Continuous Manufacturing of Solid Dose Using Direct Compaction

Abstract:

Â A Fault-tolerant Control Design for Real-time Release in Continuous Manufacturing of Solid Dose using Direct Compaction

Qinglin Su, Mariana Moreno, Jianfeng Liu, Sudarshan Ganesh, Yasasvi Bommiready, Marcial Gonzalez, Gintaras V. Reklaitis, Zoltan K. Nagy

School of Chemical Engineering, Purdue University,
West Lafayette, IN 47907

Thomas Oâ€™Connor, Tian Geng

CDER, Office of Pharmaceutical Quality, U.S. Food and Drug Administration,
Silver Spring, MD 20993

Abstract

Advantages of continuous manufacturing in the pharmaceutical industry have been demonstrated through extensive research studies in the last decade [1], but mostly at the conceptual or theoretical level. Only few continuous manufacturing processes have been approved by the US Food and Drug Administration thus far, e.g., the Vertex Orkambi and Janssen Darunavir [2]. To move from a conceptual continuous design to a practically operating pilot plant or manufacturing process, the efforts of regulators, academics, and industry are now focused more on quality control strategies [3]. Specifically, the active process control design for real-time release and its risk management in continuous manufacturing [4, 5] have been considered as critical for successful implementation of continuous manufacturing processes.

In this study, a fault-tolerant control design for real-time release is implemented, according to our previous systematic framework for control design and analysis [5], to a pilot plant at Purdue University for continuous manufacturing of solid dosage pharmaceuticals using direct compaction. First, beside the fundamental control loops provided by the equipment vendors for unit operations of feeding, blending, tableting, etc., to manipulate the critical processing parameters (CPPs), supervisory active feedback and feedforward process control strategies are designed to maintain the critical quality attributes (CQAs) based on process analytical technology (PAT) tools and final product quality testing, e.g., Sotax Auto Test 4. In the active process control design, the major system dynamics are identified with state-space models using historical operating data in the pilot plant, and a series of rational control design and analysis metrics are employed to evaluate the control performance and screen the potential design risks [5]. For example, a Relative Gain Array (RGA) analysis is used to evaluate the system interactions and select the best-performing control pairings in the feeding-blending units, where the diagonal pairing is preferred under low system frequency domain, and is risky under higher frequency as is the case when the feeder nozzle is accidently attached to the blender hopper frame.

In the proposed fault-tolerant control design, a risk mapping, assessment and planning (Risk MAP) strategy is also considered to address the potential risks associated with the above active process control design. The active process control system is designed to mitigate or avoid quality risks. However, implementation of risk reduction measures can introduce new risks into the system or increase the significance of other existing risks (ICH Q9). In this study accepted residual risks and new potential risks to product quality after the implementation of the active process control system for the continuous manufacturing process are examined. . For example, when faults occurred in the feeding-blending units, the process performance indices, e.g., process capabilities index, are below their minimum targets. Hence, for real-time release purposes, the proposed fault-tolerant control design is evaluated experimentally and in simulation mode under these risk scenarios, e.g., process disturbance, probe fouling, instrument calibration errors, etc. It is as expected that the fault-tolerant control design with advanced process systems engineering tools, such as process monitoring, data reconciliation, state estimation, etc., can mitigate those potential risks that are common causes of variations in an otherwise stable process.

Steady-state operation of the pilot-plant is also executed over longer durations to validate the proposed fault-tolerant control design. Concluding remark will also be offered, suggesting potential guidelines for achieving real-time release in continuous solid dosage manufacturing.

References

[1] M. Ierapetritou, F. Muzzio and G. Reklaitis, "Perspectives on the continuous manufacturing of powder-based pharmaceutical processes," *AIChE Journal,* vol. 62, no. 6, pp. 1846-1862, 2016.

[2] L. Yu, "Continuous Manufacturing Has a Strong Impact on Drug Quality," US Food and Drug Administration, 12 April 2016. [Online]. Available: https://blogs.fda.gov/fdavoice/index.php/2016/04/continuous-manufacturing-has-a-strong-impact-on-drug-quality/. [Accessed 11 April 2017].

[3] S. L. Lee, T. F. O'Connor, X. Yang, C. N. Cruz, S. Chatterjee, R. D. Madurawe, C. M. V. Moore, L. X. Yu and J. Woodcock, "Modernizing pharmaceutical manufacturing: from batch to continuous production," *Journal of Pharmaceutical Innovation,* vol. 10, no. 3, pp. 191-199, 2015.

[4] R. Singh, M. Ierapetritou and R. Ramachandran, "An engineering study on the enhanced control and operation of continuous manufacturing of pharmaceutical tablets via roller compaction," *International Journal of Pharmaceutics,* vol. 438, pp. 307-326, 2012.

[5] Q. Su, M. Moreno, A. Giridhar, G. V. Reklaitis and Z. K. Nagy, "A systematic framework for process control design and risk analysis in continuous pharmaceutical solid-dosage manufacturing," *Journal of Pharmaceutical Innovation,* p. in review, 2017.

496819 Intensified Crystallization of Sulfathiazole and Sulfanilamide By Direct Co-Crystal Assembly

Abstract:

Most of the studies mainly focus on the co-crystallization step rather than on the synthesis and co-crystallization together. Based on the concept of direct co-crystal assembly, we have successfully produced 1:1 co-crystals of sulfathiazole-theophylline (STZ-THE) and 1:1 co-crystals of sulfathiazole-sulfanilamide (STZ-SNM) by integrating the chemical synthesis and co-crystallization. Interestingly, the yields of the integration method were higher than the ones of normal reactive crystallization methods, and the amount of purities were also reduced. For the chemical reaction that suffered from the low yields or had the impurity which was hard to be separated, the formation of co-crystals might be an alternative strategy to improve the efficiency of purification and the value of drugs as well. The modification of loading stoichiometry of reactants to co-formers could lead to the difference of product compositions. Both STZ-THE and STZ-SNM exhibited different pH solubility profiles. In other words, co-crystallization could change the pH effect active pharmaceutical ingredient.

497068 Direct Co-Crystal Assembly from Synthesis to Co-Crystallization

Abstract:

Pharmaceutical co-crystals were assembled directly via chemical synthesis by cooling under three new strategies. The screening method for direct co-crystal assembly by a chemical reaction was developed based on liquid-assisted grinding. The ternary phase diagram including a chemical reaction and co-crystallization was established for a 2 : 1 co-crystal of benzoic acidâ€"sodium benzoate.

502452 Continuous Preparation of 1:1 Haloperidol-Maleic Acid Salt By a Novel Solvent-Free Method Using a Twin Screw Melt Extruder

Abstract:

Salts are generally prepared by acid-base reaction in relatively large volumes of organic solvents, followed by crystallization. In this study, the potential for preparing a pharmaceutical salt between haloperidol and maleic acid by a novel solvent-free method using a twin-screw melt extruder was investigated. The pH-solubility relationship between haloperidol and maleic acid in aqueous medium was first determined, which demonstrated that 1:1 salt formation between them was feasible (pH_{max} 4.8; salt solubility 4.7 mg/mL). Extrusion of a 1:1 mixture of haloperidol and maleic acid at the extruder barrel temperature of 60 °C resulted in the formation of a highly crystalline salt. The effects of operating temperature and screw configuration on salt formation were also investigated, and those two were identified as key processing parameters. Salts were also prepared by solution crystallization from ethyl acetate, liquid-assisted grinding, and heat-assisted grinding and compared with those obtained by melt extrusion by using DSC, PXRD, TGA, and optical microscopy. While similar salts were obtained by all methods, both melt extrusion and solution crystallization yielded highly crystalline materials with identical enthalpies of melting. During the pH-solubility study, a salt hydrate form was also identified, which, upon heating, converted to anhydrate similar to that obtained by other methods. There were previous reports of the formation of cocrystals, but not salts, by melt extrusion. ^{1}H NMR and single-crystal X-ray diffraction confirmed that a salt was indeed formed in the present study. The haloperidol-maleic acid salt obtained was nonhygroscopic in the moisture sorption study and converted to the hydrate form only upon mixing with water. Thus, we are reporting for the first time a relatively simple and solvent-free twin-screw melt extrusion method for the preparation of a pharmaceutical salt that provides material comparable to that obtained by solution crystallization and is amenable to continuous manufacturing and easy scale up.

497367 Intensification of Cryogenic Lithiation-Borylation Though Use of a High Heat Transfer Loop Reactor

Abstract:

Lithium-hydrogen exchange is a common chemical transformation used to functionalize molecules, specifically to add boron to phenyl rings. One challenging class of these lithiations is the case where fluorine directs the lithium to the ortho position. These ortho-fluorine lithiated phenyl species can undergo highly exothermic decomposition through benzyne intermediates and are therefore run at deep cryogenic conditions. Not only does the intermediate need to be kept cold, but the subsequent borylation step also generates a substantial amount of heat that can promote the decomposition reaction. The challenge of handling the heat generation means operating this process requires good control of the temperature and rate of heat generation. In a batch process, this means slow addition of reagents at low temperatures. Changing operation from a batch to a continuous process presents the opportunity to increase the heat transfer rate and potentially support higher temperatures and faster mixing.

A small scale cryogenic flow chemistry platform was used to initially screen this reaction in a series of two tubular reactors.(1) At this milliliter scale, the lithiation was operated at -50 °C with the lithiated intermediate directly mixed with the boron source without degradation in a simple y-mixer. The reaction was further scaled up by approximately a factor of 3 to generate gram quantities of material and faced several challenges in the mixing section with respect to heat removal.(2) These were overcome by reconfiguring the mixing zones and better controlling heat removal from the reactor. These early results indicate that further scaling up of the process using the same configuration would present insurmountable problems.

We will discuss an additional 10x scale up capable of producing kilogram quantities of material over the course of a day. This required redesigning the borylation reactor to use a loop reactor with a heat exchanger inline to remove the heat of reaction rapidly. We will also show how careful design of the lithiation reactor would produce plug-flow performance even under laminar flow at the larger scale. The envisioned production scale continuous plant would have a 100x smaller footprint and reduction in potential chemical energy contained in the unstable intermediate. This would be safer and more economical than the equivalent batch process.

References:

1. Newby, J. et al. Design and Application of a Low-Temperature Continuous Flow Chemistry Platform. Org. Process Res. Dev. 2014, 18, 1211-1220.

2. Newby, J. et al. Reconfiguration of a Continuous Flow Platform for Extended Operation: Application to a Cryogenic Fluorine-Directed ortho-Lithiation Reaction. Org. Process Res. Dev. 2014, 18, 1221-1228.

497922 Design of Novel Microalgal Photbioreactor Using Computational Fluid Dynamics

Abstract:

The incessant rise of global CO_2 levels in the environment, depleting fossil fuel reserves and increased human & industrial activities are putting our ecosystem at risk. The search of renewable and sustainable options for generating fuels are on the rise. Microalgae is regarded as a green alternative owing to its several beneficial characteristics such as faster growth rates, CO_2 fixation abilities, etc. Moreover, microalgae has a tremendous potential for efficient CO_2 capture from point sources, biomass production for bio-product generation (e.g. biofuels, cosmetics, fodder and food etc.) and waste-water mitigation. However, one of the major bottlenecks in the microalgae-to-biofuel value chain lies in the growth of microalgae for commercial scale. Microalgae is cultivated in open as well as closed systems. However, the improved performance of a microalgal growth system is generally exhibited in closed photobioreactor which is governed by several physico-chemical factors such as light, pH, mixing, aeration, CO_2 etc. It is generally true that hydrodynamics plays a significant role by improving the mixing of nutrients, gases (improved mass transfer), avoiding the creation of dead zones, improved light penetrability and reduced sheer stress on the biological cells as well as reduced power consumption. In field experiments as well as computer simulation studies, the most common types of PBRs which are thoroughly investigated are bubble column and air-lift reactors. However, there is a scope to enhance flow patterns in PBRs using advanced CFD techniques. Computational Fluid Dynamics (CFD) enables determination of the hydrodynamics of the culture medium and mixing efficiency according to various PBR designs which is otherwise difficult to measure in field experiments. In this work Eulerâ€"Lagrangian model has been used to evaluate the turbulence mixing for two phase flow in airlift PBR using commercial CFD software. Three reactor configurations have been proposed with variation in draft tube length and type for assessing its effect on hydrodynamic characteristics. Parameters such as flow rate (Q), mean velocity (**u**), and mean Turbulent Kinetic Energy (TKE) were characterized for each configuration. The dimensions of the novel design were optimized using iterative design procedure. In all experiments, tap water was used as the liquid phase and was operated in a batch mode, while compressed air (CO_2 rich) was used as the gas phase and was operated in continuous mode. The hydrodynamics properties such as axial dispersion coefficient, mixing time and circulation time were calculated. The experimental results and CFD simulations were thoroughly compared for the novel PBR with that of conventional airlift reactor. In addition to performance evaluation of the novel PBR, a green microalga *Chlorella* sp. is cultivated in the PBR using air/flue gas and its growth characteristics were determined. The performance of the novel PBR was compared with that of the conventional airlift PBR. The results showed that the novel design had superior liquid circulation properties that resulted in a regular and alternative exposure to light and dark regions of the PBR. Also, CFD led to improved CO_2 mass transfer efficiency, biomass production, nutrient and light utilization profile.

501874 Process Intensified and Direct Production of Gasoline from Syngas

Abstract:

An Abstract Submission to 2017 AICHE Meeting

Session: Process Intensification By Enhanced Heat and Mass Transfer

Topical Area: Process Intensification

Process Intensified and Direct Production of Gasoline from Syngas

Xinquan Cheng[a], William R. Yantz Jr.[a], and Bruce J. Tatarchuk[a, b,*]

[a] Center *for Microfibrous Materials Manufacturing, Department of Chemical Engineering, 212 Ross Hall, Auburn University, AL 36849*

[b] *IntraMicron Inc., 368 Industry Dr., Auburn, AL 36832, USA*

* *Corresponding author. Tel.: +1 334 844 2023; fax: +1 334 844 2065; E-mail address: Tatarbj@auburn.edu*

Abstract

The increasing demand for liquid fuels draw people's attentions to the synthetic fuels. Currently, there are mainly two approaches to synthesize liquid fuels from syngas. One is Fischer-Tropsch synthesis (FTS) and another one is syngas-to-methanol first and followed up with methanol-to-gasoline (MTG). Those two approaches are well-known techniques for converting syngas to liquid fuels. However, FTS process produces a broad spectrum of products, which requires complex and expensive subsequent processes to upgrade the initial products in order to meet liquid fuel specifications. The MTG approach is a multiple step process. Typically, in the first reactor syngas is converted to methanol, and followed with methanol dehydration reactor to produce the intermediate product dimethoxyethane (DME), and then DME successively converted to hydrocarbons in the reactor packed with zeolite. It must be noticed that all those reactions are highly exothermic reactions, which producing extremely amount of heat during the reaction processes. Therefore, the reaction thermal management is always a big issue for syngas to liquid fuels.

Here, we present a single reactor system to directly convert syngas-to-gasoline (STG) by using a novel catalyst structure called Microfibrous Entrapped Catalyst (MFEC) structure. The MFEC structure was developed by our group, which is a catalyst network supported by micron-sized metal fibers with high thermal conductivity. By supporting small catalyst particles into a high voidage matrix of microfibers the catalyst is suitable for fixed bed applications without severe pressure drop or bed channeling encountered in small particle packed beds. Previous efforts on this unique catalyst structure have shown significant enhancement in intra-bed heat transfer and mass transfer.[1-4] Basically, we packed a large tubular reactor (34.0 mm I.D.) in three zones by using Cu MFEC. As we can see from Figure 1, the first zone (Zone-1) is a mixed catalyst zone for the production of DME from syngas. The second zone (Zone-2) is a relative inert zone mainly used to separate Zone-1 from Zone-3 since those two zones have different temperature. In order to convert all the unreacted methanol from Zone-1, we packed γ-Al_2O_3 in Zone-2 as a methanol dehydration catalyst. The third zone (Zone-3) is a zeolite zone packed with ZSM-5 to convert the DME to gasoline. This kind of multi-zone reactor design can separate the zeolite from the methanol synthesis catalyst, because the proximity of those two catalyst components is crucial factor for the product selectivity of STG process.[5] If those two components are close to each other, the lower olefins, which are the essential intermediates from methanol to aromatics, diffuse from the zeolite to the methanol synthesis catalyst, where they are easily hydrogenated to lower paraffins. It turns out that most of the products are short chain paraffins.[6]

However, with this simple multi-zone reactor design, the olefinic intermediates are preserved for subsequently oligomerization reactions. Under a typical reaction conditions (P=30 atm, GHSV=1000 h^{-1}, H_2:CO=1:1), the $C_5 - C_{10}$ hydrocarbon selectivity can reach 39.0 wt% with a relative high CO conversion (73.2 %). In addition, the Cu MFEC structure demonstrated a very low radial temperature gradient (around 5 °C) for a 34.0 mm I.D. reactor and insignificant amount of pressure drop. While with the same test conditions, the comparative packed bed reached a very high radial temperature gradient (around 140 °C) and significant amount of pressure drop. Considering these attributes and advantages, the Cu MFEC multi-zone reactor system is a very efficient process intensification technique for the direct production of gasoline from syngas.

Figure 1. The STG reactor design (all the catalysts particles entrapped in Cu MFEC in Zone-1, Zone-2, and Zone-3 are in 60 – 80 mesh)

References:

[1] M. Sheng, H.Y. Yang, D.R. Cahela, B.J. Tatarchuk, J. Catal. 281 (2011) 254-262.

[2] M. Sheng, H.Y. Yang, D.R. Cahela, W.R. Yantz Jr., C.F. Gonzalez, B.J. Tatarchuk, Appl. Catal. A-Gen. 445-446 (2012) 143-152.

[3] M. Sheng, D.R. Cahela, H.Y. Yang, C.F. Gonzalez, W.R. Yantz Jr., D.K. Harris, B.J. Tatarchuk, Int. J. Heat Mass Tran. 56 (2012) 10-19.

[4] X.Q. Cheng, H.Y. Yang, B.J. Tatarchuk, Catal. Today 273 (2016) 62-71.

[5] S. Sartipi, M. Makkee, F. Kapteijn, J. Gascon, Catal. Sci. Technol. 4 (2014) 893-907.

[6] K. Fujimoto, H. Saima, H. Tominaga, J. Catal. 94, (1985) 16-23.

487118 Modelling of Time-Dependent Interfacial Properties Due to Chemical Equilibrium Reactions in Demixed Fluid Systems

Abstract:

Modelling of time-dependent interfacial properties due to chemical equilibrium reactions in demixed fluid systems

Andreas Danzer, *Sabine Enders*

Karlsruhe Institute of Technology, Institute of Technical Thermodynamics and Refrigeration Engineering, Engler-Bunte-Ring 21, D-76131 Karlsruhe, Germany.

Email: andreas.danzer@kit.edu; sabine.enders@kit.edu

Key words: reactive extraction, impact of chemical reactions on interfacial properties, model calculations

Abstract:

Reactive extraction, the integration of reaction and extraction within one unit operation, is a promising process in a variety of applications. Compared with the conventional process of carrying out reaction and separation sequentially, the in-situ removal of products formed in the reaction zone can favorably influence conversion and selectivity, especially of equilibrium reactions. Together with a significantly reduced effort in the task of separation, this can result in major energy as well as capital cost savings. However, there are feasibility constraints resulting both from thermodynamics and chemistry. In addition, the process intensification leads to higher complexity and thus the need for research in terms of modelling productiveness. Because of that model calculations of hypothetically ternary liquid-liquid reactive extraction systems were performed in this work. The liquid-liquid phase equilibrium was calculated with the well-known, quite simple Margules activity model, the interfacial properties were modelled with the Margules equation in combination with the Density Gradient Theory [1-4] (DGT). It was assumed that the chemical reaction occurs in a A B manner, whereas the solvent C does not participate in the chemical equilibrium reaction. Furthermore, the assumption was made that mass transfer occurs at speed of infinity, so the chemical reaction is rate determining, in contrast to the reach of phase equilibrium. The empirical constants of the Margules equation $A_{i,j}$, the so called influence parameter κ of the DGT as well as the chemical equilibrium constant $K=k_1/k_2$ were specified due to the need to serve as a model. For real systems, $A_{i,j}$ must be fitted to experimental equilibrium data of the binary subsystems, the influence parameter κ has to be fitted at one experimental data point of the interfacial tension of each binary system with a miscibility gap and K is determined by the chemical reaction. It was found that the time, to reach chemical equilibrium in addition to the phase equilibrium, is strongly dependent on the shape of the miscibility gap as well as on the slope of the tie lines. Furthermore, the interfacial tension of the demixed system decreases over time, because of the fact that the product of the chemical equilibrium reaction act as a solubilizer. The slope of the interfacial tension over time strongly corresponds to the quotient of the reaction rate of forward and backward reaction. The new theoretical framework allows to the first time the calculation of time-dependent interfacial tensions and can further transferred to real systems.

Literature:

[1] J.W. Cahn, J.E. Hilliard, J. Chem. Phys. 28 (1958) 258.

[2] C.I. Poser, I.C. Sanchez, Macromolecules 14 (1981) 361.

[3] T. Grunert, H. Rudolph, S. Enders, Zeitschrift für Physikalische Chemie 227 (2013) 1.

[4] T. Grunert, S. Enders, Fluid Phase Equilibria 381 (2014) 46.

495073 Impact of Internals on Bubble Column Performance

Abstract:

Impact of internals in a bubble column reactor is a widely studied topic especially for the interesting hydrodynamics involved. Though numerous experiments have been performed utilizing horizontal, vertical internals and other packing materials, their direct, individual impacts on the bubble column performance is complex and not fully understood. Utilizing effective CFD simulations to recreate the flow conditions has become a necessity to understand these dynamics better. Based on the reaction, operating regime and flow conditions, it is possible to drive the back mixing, liquid recirculation and other parameters favorable to our requirements, by modifying the internal structures. We made an attempt to quantify the performance of various internal designs (taken from literature), for a given column, based on selected hydrodynamic parameters. With a better understanding of the flow patterns and their response to a number of internal structures, we take this forward and propose new designs which are tailored specifically for the reaction conditions and maintaining the flow in a particular regime (bubbly flow regime in our case) even at higher U_g thereby successfully delaying transition. We validate them by 3-d printing these designs and testing out their performance on a laboratory scale bubble column.

501007 Electrospun Collagen Scaffold for Peripheric Nerve Regeneration

Abstract:

Damage of peripheral nervous system directly interferes with the communication network between brain, spinal cord and the rest of the body, and around 20 million people in United States suffer this [1]. Donor site morbidity and limited graft material are limitations confronted by solutions such as allograft and autograft [2]. Otherwise nerve conduits have the capacity to develop a successful nerve recovery , and most important return a better quality of life to patients. The lost of extracellular matrix (ECM), local growth factors and cells is what limits the regeneration of nerves.The proposed research consists of the engineer of 3D collagen nanofiber scaffold through electrospinning, a versatile technique for nanofibers development . The scaffold, composed of the most abundant protein in our body collagen, mimics ECM and facilitates cell adhesion and elongation to promote nerve regeneration, providing biological and mechanical support[3][4]. The development of the collagen nanofiber structure is optimized via electrospinning parameters in order to modify nanofiber orientation which is directly related towards the direction of nerve regeneration. A random and aligned pattern has been developed and confirmed by Scanning Electron Microscope (SEM). In vitro assays and fluorescene microscope analysis with Schwann and PC-12 cells, both key cell lines in nerve tissue regeneration, resulted on successful cell adhesion to electrospun collagen nanofiber, random and aligned. Cellular tissue has been observed on collagen scaffold through SEM, after 24h cell culture and proper cell fixation method. Present experiments consist on the addition of nerve growth factor (NGF) to PC-12 cell line in order to promote differentiation. The differentiation of PC-12 cells will be analyzed towards the influence of collagen scaffold orientation on the direction of cell growth elongation. A future approach is a co-culture between Schwann and PC-12 ,on collagen scaffold,with the vision to prove that the naturally secreated NGF by Schwann cells promote the differentiation of PC-12 . Results from this co-culture will validate the use of collagen electrospun nanofiber scaffold impregnated with Schwann cells as a vehicle to deliver the proper natural components in a injured nerve and achieve nerve regeneration.

References:

1. National Institute of Neurological Disorders and Stroke: Peripheral Neuropathy Fact Sheet (Bethesda, Maryland, 2014).

2. W. Z. Ray and S. E. Mackinnon: Management of nerve gaps: Autografts, allografts, nerve transfers, and end-to-side neurorrhaphy. *Exp. Neurol.* **223**(1), 77 (2010).

3. S. Previtali, M. Malaguti, N. Riva, M. Scarlato, P. Dacci, G. Dina, D. Triolo, E. Porrello, I. Lorenzetti, R. Fazio, G. Comi, A. Bolino, and A. Quattrini: The extracellular matrix affects axonal regeneration in peripheral neuropathies. *Neurology* **71**(322–331) (2008).

4. G. a. Di Lullo, S. M. Sweeney, J. Körkkö, L. Ala-Kokko, and J. D. San Antonio, "Mapping the ligand-binding sites and disease-associated mutations on the most abundant protein in the human, type I collagen," *J. Biol. Chem.*, vol. 277, no. 6, pp. 4223–4231, (2002).

501165 3D Printing of Nerve Guidance Channels for Peripheral Nerve Repair

Abstract:

The peripheral nervous system (PNS) is a complicated and extensive network of nerves that are the means by which the brain and spinal cord control the rest of the body. The PNS is fragile and can be easily damaged by injuries or trauma. Surgical treatment is the only remedy currently available, with the gold standard for defects greater than 8 mm being autologous nerve grafts; however only around 40% of the 1.8 million US PNS patients each year regain normal function. In addition, nerve grafts have been particularly ineffective at repairing critical-size nerve defects (> 3 cm).[1] Scaffold-based strategies where a tubular nerve guidance channel (NGC) is used to bridge the nerve defect have been promoted as a potential alternative that could avoid the additional surgeries and associated donor site morbidity involved in the harvest of nerve grafts.[2] Clinicians have thus increased the use of NGCs combined with current surgical therapeutics. However, current NGCs lack patient-specific tunability and are only approved for small-gap (< 3cm) injuries by the U.S. Food and Drug Administration (FDA). Current research efforts are focused on creating more complex NGCs that can support regeneration of critical-size defects.

In this context, our research seeks to use additive manufacturing technologies to create bioactive and cellular NGCs on demand for the repair of critical-size nerve defects. Recently, 3D printing has been increasingly used in research and medical therapeutics for rational, computer-aided design of biomaterial-based scaffolds with complex architecture. Furthermore, printing with co-axial extruders can enable the direct printing of layered tubular structures for use as NGCs. The NGCs should contain an outer flexible shell that seeks to mimic the mechanical properties of the surrounding biological tissue and enable diffusion of nutrients to support encapsulated cells. The use of biodegradable block copolymers with both hydrophilic and relative hydrophobic functions can provide a flexible, partially-hydrated, biocompatible and bioresorbable NGC shell.

In this study, an A-B type diblock copolymer of poly(ethylene glycol) methyl ether (mPEG) combined with poly(l-lactide) acid (PLLA) and another A-B-A type triblock copolymer of PLLA-PEG-PLLA were synthesized using varied ratios of mPEG, PEG and PLLA. The resulting block copolymers of mPEG-PLLA and PLLA-PEG-PLLA containing different lengths of the PLLA hydrophobic chain were characterized with gel permeation chromatography (GPC), differential scanning calorimetry (DSC), and nuclear magnetic resonance (NMR) to determine molecular weight, polymer structure, and thermal behavior. In addition, equilibrium water content, degradation rates, and cell response were all evaluated and correlated to polymer structure. With the 3D printing of copolymers using coaxial printing needles, a scalable process for the production of cellular NGCs could be developed.

References

1. Burkhard Schlosshauer, L.D., et al., Neurosurgery, 59 (2006), 740-748.

2. Cinteza, D., et al., Maedica â€“ A Journal of Clinical Medicine, 10:1 (2015), 65-68.

489637 Combined Physical and Biochemical Cues Direct the Growth of Inner Ear Neurites

Abstract:

Cochlear implant (CI) devices help to restore hearing to individuals who suffer from sensorineural hearing loss. While CIs provide basic auditory perception, implant patients only experience limited voice and tonal perception. Due to spatial separation between the electrode array and the neural receptors, significant current spread and nonspecific stimulation of the spiral ganglion neurons (SGNs), or inner ear neurons, occurs. Thus, devices with advanced electrical design are unable to recapitulate the tonotopic stimulation of neurons because of spatial resolution. Strategies to decrease the distance between CI electrodes and primary neural receptors have focused on regenerating the SGN neurites. To be effective, the spatial order of the neurites must remain intact and thus cellular cues must orient and guide the pathfinding of SGN neurites. In this work, physical and biochemical cues are used to direct the regrowth of SGN neurites. Utilizing the inherent temporal and spatial control of photopolymerization, physical microgrooves are fabricated using a photomask in a single step process. The spatial control of the photon-induced reaction forms ridges under transparent band and depressions under reflective regions with channel amplitudes ranging from 250 nm to 10 μm. Biochemical patterns are generated by adsorbing laminin, a cell adhesion protein, to acrylate polymer surfaces followed by irradiation through a photomask with UV light to deactivate protein in exposed areas and generate parallel biochemical patterns. Laminin showed deactivation with increased exposure time as measured by immunofluorescence. Adding photoinitiator decreased the time required to deactivate the adsorbed laminin. To evaluate neural response to the patterns, SGNs were first cultured on physically patterned substrates and showed increased alignment with higher amplitude and lower periodicity polymers. SGNs were also cultured on biochemical patterns and showed excellent alignment to the parallel bands confirming the photodeactivation process was effective in creating active laminin patterns. To further evaluate the ability of biochemical and physical features to be used in overcoming conflicting cellular cues, competing patterns were tested. Physical microgroove channels were first fabricated followed by a perpendicular laminin protein pattern. The strength of the physical cue was independently varied by changing the amplitude or the periodicity. Lower periodicity, or more frequent, surfaces induced a greater degree of alignment along the channels. Higher amplitudes, or deeper channel depths, also increased the guidance of the SGNs. When SGN neurites were seeded onto substrates with perpendicular biochemical and physical patterns, alignment was strongly dependent on both amplitude and periodicity. At low amplitudes neurites tended to grow along the laminin stripes. As the amplitude was decreased, alignment was disrupted and neurite segments followed neither the biochemical or physical pattern. Similarly, at lower periodicity, SGN neurites were not strongly aligned to either pattern. As the periodicity was increased, alignment shifted to follow the biochemical laminin pattern. These results demonstrate the ability of photopolymerized micro-features to modulate alignment of inner ear neurites even in the presence of conflicting physical and biochemical cues laying the groundwork for next generation cochlear implants and neural prosthetic devices.

495713 Harnessing Multi-Functional Microbial Cells for Designing Sweat-Responsive Bio-Hybrid Wearables

Abstract:

We found a variety of microbial cells and their cellular components have hygroscopic behavior. By depositing these biomaterials on top of a humidity-inert film, a multi-layered bio-hybrid material is obtained, which can reversibly change shape and biofluorescence intensity in response to environmental humidity change. We characterized the performance of the film, which guide the design of a wearable running suit and a fluorescent shoe prototype. Those bio-flaps on wearables can dynamically modulate ventilation in synergy with the bodyâ€™s need for cooling.

497286 Development of Silk Protein Conjugates for Mucoadhesive Applications

Abstract:

Polymers that adhere to wet surfaces that are biocompatible, degradable, and devoid of animal-sourced materials are desired for applications as tissue adhesives. Silk from the domesticated silkworm *Bombyx mori* is a naturally occurring biopolymer with charged hydrophilic terminal regions that endcap a blocky core dominated by hydrophobic domains. Silk was selected for modification for use as an aqueous adhesive because of its high molecular weight, tunable mechanical and degradation properties, aqueous processability, and for its hydrophobicity that was envisioned to afford control over the mechanical properties and aqueous swelling of the modified biopolymers. This work sought to exploit the unique structure and properties of this protein by synthesizing gels that are stable in an aqueous environment and adhere to mucus-covered surfaces. To prepare adhesive hydrogel coatings that can be applied as a liquid and subsequently gel, silk was successfully conjugated with tyramine, a phenol-containing compound capable of being oxidized to a catechol functional group using mushroom tyrosinase (MT). As both aqueous and organic phase syntheses are possible with silk, several solvent systems were investigated to determine the conditions that led to successful coupling without protein aggregation. The silk polymers were characterized to determine composition, adhesive strength using lap shear measurements, cell compatibility and support of proliferation, and their ability to form beta sheets, which physically crosslink silk. Importantly, the conjugation of tyramine showed no effect on silkâ €™s ability to form beta sheets, which may be used to further tune the materialâ€™s mechanical properties and degradation profile. These materials afford a new route to synthesize silk-based gels, which we envision to have utility as biodegradable adhesives and sealants.

499466 Laser-Activated Sealants for Skin Tissue Repair

Abstract:

Sutures and staples are ubiquitous in surgery, but present their own complications and cause local trauma to tissue during application. Glues, adhesives and sealants have been investigated as means for approximating wounds, but many suffer from toxicity and poor strength. Photothermal tissue sealing and repair, facilitated by near-infrared laser-responsive nanoglues, has emerged as a potential alternative to existing tissue apposition methods. Here, we demonstrate the photothermal sealing of lacerted skin and other ruptured tissues using laser activation of novel nanoglues in which, gold nanorods (GNRs) are embedded throughout a polypeptide matrix. Critical factors including laser power density, exposure time, and laser type (continuous wave or pulsed wave) were investigated in order to determine optimal treatment parameters that led to the highest recovery in sealing strength. Laser-activated sealing of full thickness lacerations in skin of live mice resulted recovery of tissue mechanical properties and integrity that were comparable to that seen with commercial sutures. Local inflammation caused by photothermal sealing approach was minimal and was compared to that seen with commercial sutures. Furthermore, laser-activated sealing was also employed to reinforce suture lines in skin, as a combination approach that can provide significant benefit for rapid sealing and healing of skin lacerations. Our results demonstrate that polypeptide-GNR nanoglues may have applications in photothermal sealing and repair of ruptured tissues and may emerge as attractive alternatives to sutures.

497772 Layer-By-Layer Assembled Thin Film Biomaterials As Porous Biomolecular Delivery Systems

Abstract:

Biomaterials capable of delivering controlled quantities of bioactive agents, while maintaining mechanical integrity, are needed for a variety of cell contacting applications. We describe here two strategies toward porous, polyelectrolyte-based thin films capable of controlled biomolecular loading and release. In one approach, termed "nanoparticle templating," films are formed via the layer-by-layer assembly of charged polymers and nanoparticles (NP), then chemically cross-linked to increase mechanical rigidity and stability, and finally exposed to tetrahydrofuran to dissolve the NP and create an intra-film porous network. In another approach, termed "solution shock," films are subjected briefly to an extreme acidic environment following LbL assembly, causing expansion and ultimately a spinodal-like breakup of the film into a porous structure. We report here on film structure and mechanics, on biomolecular agent loading and release, and on the behavior of cells cultured on these films. While both strategies yield mechanics and biomolecular agent loading that are suitable for applications such as tissue engineering, the solution shock method offers advantages in simplicity and controllability.

499423 Oral Formulations: Perspectives from Additive Manufacturing

Abstract:

Selection of oral formulation strategy plays a significant role in development of new drug candidates throughout the development and commercial product stages; requirement for oral bioavailability must be balanced against additional complexity in the manufacturing process[1,2]. Formulation strategies may present manufacturing challenges, particularly for poorly water-soluble drugs, resulting from the necessity of particle size reduction and multiple active excipients to achieve bioavailability targets[3]. As an alternative to established manufacturing strategies, much recent work has focused on additive manufacturing, partially due to its capacity for accurate and precise delivery of drug amounts in the micro- to milli-gram range[4]. The technology also exhibits potential as a flexible platform capable of producing doses in a number of forms from solvent- and melt-based forumulations[5,6], which may include colloidal particles[7].

The manufacture of multiple capsule-based formulations using non-colloidal particulate suspensions is presented in this study. Particles selected from micronized and non-micronized powder grades are characterized in terms of particle size and bulk powder properties, which, in concert with rheology experiments are used to guide printing fluid formulations. Quantitative measures of product physical properties and respective production rates are presented for four active substances in 1-100 milligram amounts, delivered by hydro-philic/phobic carrier fluid platforms. Based on these results, perspectives are given on the ability of the platforms to complement current formulation selection strategies for BCS Class I and Class II drugs and on the potential for the technology serve at both the personalized medicine and industrial manufacturing scales.

References

1. M. Kuentz, R. Holm, and D. P. Elder, 2016, Methodology of oral formulation selection in the pharmaceutical industry, European Journal of Pharmaceutical Sciences, 87, 136-163.
2. M. Ierapetritou, F. Muzzio, and G. Reklaitis, 2016, Perspectives on the Continuous Manufacturing of Powder-Based Pharmaceutical Processes, AIChE Journal, 62, 1846-1862.
3. H.D. Williams, N.L. Trevaskis, S.A. Charman, R.M. Shanker, W.N. Charman, C.W. Pouton, and C.J.H. Porter, 2013, Strategies to Address Low Drug Solubility in Discovery and Development, Pharmacoligical Reviews, 65, 315-499.
4. R. Daly, T.S. Harrington, G.D. Martin, and I.M. Hutchings, 2015, Inkjet printing for pharmaceutics â€" A review of research and manufacturing, International Journal of Pharmaceutics, 494, 554-567.
5. L. Hirshfield, A. Giridhar, L.S. Taylor, M.T. Harris, and G.V. Reklaitis, 2014, Dropwise Additive Manufacturing of Pharmaceutical Products for Solvent-Based Dosage Forms, Journal of Pharmaceutical Sciences, 103, 496-506.
6. E. IÃ§ten, A. Giridhar, L.S. Taylor, Z.K. Nagy, and G.V. Reklaitis, 2015, Dropwise Additive Manufacturing of Pharmaceutical Products for Melt-Based Dosage Forms, Journal of Pharmaceutical Sciences, 104, 1641-1649.
7. J. Pardeike, D.M. Strohmeier, N. Schrödl, C. Voura, M. Gruber, J.G. Khinast, and A. Zimmer, Nanosuspensions as advanced printing ink for accurate dosing of poorly soluble drugs in personalized medicines, 2011, International Journal of Pharmaceutics, 420, 93-100.

487351 Predicting the Free Energy of Solvation: A Hybrid QSPR Model for Organic Solute/Solvent Pairs

Abstract:

The free energy of solvation is a useful quantity in the journey of a pharmaceutical molecule from discovery through to commercial process development and product design. In this paper, we review some of the roles played by this property across the moleculeâ€™s life cycle, including in the early assessment of the quantities such as the octanol/water partition coefficient or on the assessment of solvent effects on reaction energetics for screening purposes. Initial efforts to propose quantitative structure property relationships (QSPR) and predict the Gibbs free energy of solvation for a wide range of solute/solvent pairs are presented here. They include the use of partial least square (PLS) and multivariate linear regression (MLR) methods. A dataset consisting of 303 solutes in 215 solvents with total number of 1800 Gibbs free energy data points has been built. The proposed models combine experimental descriptors with quantum mechanical (QM) descriptors. For the solvents, experimental descriptors are proposed, as these provide a good description of the bulk properties of solvents and the range of solvents considered is usually relatively narrow; twelve of these descriptors are used in the proposed model (surface tension, dipole moment, octanol-water partition coefficient, dielectric constant, refractive index, boiling point, molecular weight, enthalpy of vaporization, molar volume, critical temperature, critical pressure, and critical volume). For the solutes, on the other hand, QM descriptors are used, making it possible to model solutes that have not been synthesized or that are short-lived/unstable; nine QM solute descriptors are proposed (polarizability, energy of the highest occupied molecular orbital, energy of the lowest unoccupied molecular orbital, dipole moment, most positive partial charge on hydrogen atom in molecule, most negative atomic partial charge in molecule, molecular van der Waals volume, total energy at 0 K, and entropy at 298 K). The accuracy of these initial models is found to be acceptable for fast calculation of the Gibbs free energy of solvation of a wide range of solutes in different solvents and to compare favourably with existing methods. Building on these encouraging results further work to explore whether this level of accuracy is suitable for a range of activities, including screening and process design will also be of interest.

492658 *In silico* Prediction of Absolute Rates of Crystal Growth from Solution

Abstract:

Predicting rates of crystal growth from solution *in silico* remains extremely challenging, but presents a lucrative engineering opportunity to more efficiently design crystals for pharmaceutical, catalytic, and other purposes. Mechanistic crystal growth models [1-3] capture key thermodynamic and kinetic phenomena, and they have successfully predicted steady-state crystal shapes. [4,5] These models can facilitate *in silico* crystal growth rate predictions if rates of attachment to and detachment from kink sites are known; kinks are favorable surface sites for attachment of growth units (atoms, molecules, etc.). Complex computational techniques are currently the only approach to calculate these rates. Moreover, these methods are necessary to gain mechanistic insight to aid future development of a model to predict attachment/detachment rates.

In this work, we demonstrate that *in silico* predictions of absolute crystal growth rates can yield favorable agreement with experimentally measured growth rates. We examine a proof-of-concept system â€" NaCl in water â€" which provides valuable insights that will aid future investigations of organic molecules and organic salts. We compute the rates of Na^+ and Cl^- attachment and detachment at kink sites and also elucidate the mechanistic role of water molecules during these events. [6] These rates are implemented in the mechanistic spiral growth model [7,8] to make *in silico* NaCl crystal growth rate predictions, which are in approximate agreement with experimental measurements. [9-11] These results promote the investigation of additional systems and, ultimately, the development of a model to predict attachment and detachment rates. It will be interesting to explore the role of rotational degrees of freedom and functional groups during kink attachment/detachment of organic molecules.

[1]Â J. Li, C.J. Tilbury, S.H. Kim, M.F. Doherty, *Prog. Mater. Sci.* **2016**, *82*, 1-38.

[2]Â M.A. Lovette, A.R. Browning, D.W. Griffin, J.P. Sizemore, R.C. Snyder, M.F. Doherty, *Ind. Eng. Chem. Res.* **2008**, *47*, 9812-9833.

[3] P. Dandekar, Z.B. Kuvadia, M.F. Doherty, *Annu. Rev. Mater. Res.* **2013**, *43*, 359-386.

[4] S.H. Kim, P. Dandekar, M.A. Lovette, M.F. Doherty, *Cryst. Growth Des.* **2014**, *14*, 2460-2467.

[5] C.J. Tilbury, D.A. Green, W.J. Marshall, M.F. Doherty, *Cryst. Growth Des.* **2016**, *16*, 2590-2604.

[6] M.N. Joswiak, M.F. Doherty, B. Peters, *in prep.*

[7] W.K. Burton, N. Cabrera, F.C. Frank, *Phil. Trans. R. Soc. Lond. A* **1951**, *243*, 299-358.

[8] R. Snyder, M.F. Doherty, *P. Roy. Soc. A-Math. Phy.* **2009**, *465*, 1145-1171.

[9] H. Offerman, G. von Brachel, A. Al-Sabbagh, F. Farelo, *Cryst. Res. Technol.* **1995**, *30*, 651-658.

[10] S.B. Zhang, J.J. Yuan, H.A. Mohameed, J. Ulrich, *Cryst. Res. Technol.* **1996**, *31*, 19-25.

[11] S. Al-Jibbouri, J. Ulrich, *J. Cryst. Growth* **2002**, *234*, 237-246.

501156 Mathematical Modeling and Simulation of Magnetophoresis. Application to Drug Delivery Using Magnetic-Field

Abstract:

The effects of magnetic field on paramagnetic species such as ions, biological molecules and solid particles have been investigated theoretically[1-4] and experimentally.[5,6] These paramagnetic species in the presence of an applied magnetic field experiences three kinds of body forces such as paramagnetic gradient force due to concentration gradient, field gradient force due to magnetic field gradient, and Lorentz force due to interaction of moving charged species with the magnetic field.[3,7] The flux of a population of paramagnetic species due to these magnetic body forces have been previously modeled using four different approaches- (A) the magnetic forces are included as the external body forces in the Navier-Stokes equation and the resulting velocity is added to the convective term of the flux expression,[2,3,8] (B) the velocity of paramagnetic particles is approximated as a product of mobility and a net magnetic force, which is then included in the convective term of the flux expression,[4,9,10] (C) the mobility in the migration term of the Nernst-Planck equation is modeled as a matrix containing Hall and Transversal mobilities, which are the functions of magnetic fields,[1] and (D) the diffusion coefficients are modified to include the effects from Lorentz force and the velocity in the convective term is modeled as a product of mobility and field gradient force.[11] The approaches A-D have a few common limitations that these models are not applicable to the concentrated solutions of paramagnetic particles, and they do not account for all magnetic forces including magnetic dipole interactions between particles. Here we propose a novel model for the flux of paramagnetic species that overcomes all the limitations of the preceding approaches. Moreover, the proposed model is based on the first principles that accounts for fluxes due all three magnetic forces with inter-particle interactions.

References:

(1) Aaboubi, O.; Chopart, J.; Douglade, J.; Olivier, A.; Gabrielli, C.; Tribollet, B. Journal of the Electrochemical Society 1990, 137, 1796.

(2) Waskaas, M.; Kharkats, Y. I. The Journal of Physical Chemistry B 1999, 103, 4876.

(3) Hinds, G.; Coey, J.; Lyons, M. Electrochemistry communications 2001, 3, 215.

(4) Furlani, E. Journal of Applied Physics 2006, 99, 024912.

(5) Lueshen, E.; Venugopal, I.; Soni, T.; Alaraj, A.; Linninger, A. Journal of biomedical nanotechnology 2015, 11, 253.

(6) Venugopal, I.; Pernal, S.; Fusinatto, T.; Ashkenaz, D.; Linninger, A. Nano Biomedicine & Engineering 2016, 8.

(7) Weston, M. C.; Gerner, M. D.; Fritsch, I. Analytical chemistry 2010, 82, 3411.

(8) Waskaas, M.; Kharkats, Y. I. Journal of Electroanalytical Chemistry 2001, 502, 51.

(9) Furlani, E.; Ng, K. Physical review E 2006, 73, 061919.

(10) Furlani, E. P.; Xue, X. Microfluidics and nanofluidics 2012, 13, 589.

(11) Kinouchi, Y.; Tanimoto, S.; Ushita, T.; Sato, K.; Yamaguchi, H.; Miyamoto, H. Bioelectromagnetics 1988, 9, 159.

486626 Polyalkoxylated Alcohols As Excipients for Poorly Soluble Drugs

Abstract:

Polyalkoxylated Alcohols as Excipients for Poorly Soluble Drugs

Thomas H. Kalantar[1], Jin Zhao[2], Mladen Ladika[1], Keith Harris[1], Chris Tucker[1], Michael Tulchinsky[1], TC Kuo[1], Joe Kiefer[1], Xiaoyun Chen[1], Robert Krystosek[1]

(1) The Dow Chemical Company, Herbert D. Doan R&D Center, Midland, Michigan, 48674, USA

(2) Pharmaceutical Excipients R&D, Dow Food, Pharma & Medical Films, Midland, MI 48674

There has been a steady increase in the number of active pharmaceutical ingredients (APIs) that are characterized with low aqueous solubility. It has been estimated that 40 to 80% of new drug candidates suffer from poor aqueous solubility and/or slow dissolution, reducing their bioavailability and efficacy, and even preventing their general use [1]. This study explored a solubilization approach using low-T_g alkylated polyoxyalkylene copolymers to enhance drug solubility and dissolution rate[2], where the drug was combined with the polymeric excipient to form an amorphous solid dispersion at the molecular level. Such low-T_g excipients yield liquid or semi-solid drug/excipient formulations.

In this study, a novel class of alkylated polyols was synthesized using renewably sourced alkyl initiators, with a specific emphasis on terpene alcohol precursors. These alcohols were first alkoxylated with butylene oxide then further ethoxylated with ethylene oxide. The effect of solubilization by these copolymers was studied using four model drugs (ketoprofen, danazol, griseofulvin, and probucol). Raman spectroscopy suggested that several copolymers from this study were effective in stabilizing ketoprofen in its amorphous form. The aqueous solubility and dissolution rate of drugs in presence of these excipients were studied using nephelometry and optical microscopy. Several drug/excipient samples showed a dramatic reduction in turbidity relative to the turbidity of drugs alone in water. Optical microscopy of those drug/excipient samples in water also indicated an increase in drug dissolution. Drug solubility was also measured by an HPLC method in which a capsule filled with probucol/excipient mixture was placed in phosphate buffer solution for 30 minutes after capsules were disintegrated.

Medium to high enhancement of solubility were observed with blends of common commercial polymeric excipients with three drugs: phenytoin, griseofulvin, and probucol. Probucol solubility was significantly enhanced when mixed with excipient polymers at a molecular level. When mixtures of probucol and novel alkyl-BO-EO excipients were loaded into capsules and subjected to ambient conditions for one week and accelerated aging at 40 °C and 75% RH for 4 months, the observed increases in probucol solubility were significant (up to 8,000-fold) under both sets of storage condition.

Many polyoxyalkylene copolymers can be used as solubilization-excipients for poorly soluble drugs; however the structure of alkyl-BO-EO copolymers can be optimized to specific drug structures to enable otpimization of solubilization.

References

Â

1. Benet, L. Z., The role of BCS (biopharmaceutics classiï¬cation system) and BDDCS (biopharmaceutics drug disposition classiï¬cation system) in drug development. *J Pharm Sci.* **2013,** *102* (1), 34-42.

2. (a) Tang, J.-L.; Sun, J.; He, Z.-G., *Current Drug Therapy* **2007,** *2*, 85-93; (b) Nielsen, F. S.; Petersen, K. B.; Mullertz, A., *Eur. J. Pharm. Biopharm.* **2008,** *69*, 553-562; (c) Kohli, K.; Chopra, S.; Dhar, D.; Arora, S.; Khar, R. K., *Drug Dicsovery Today* **2010,** *15*, 958-965; (d) Holm, R.; Jorgensen, E. B.; Harborg, M.; Larsen, R.; Holm, P.; Mullertz, A.; Jacobsen, J., *Eur. J. Pharm. Sci.* **2011,** *42*, 416-422.

503694 2D Printed Product Design of Patient Centric Dosage Forms for Adaptive Clinical Trials

Abstract:

Adaptive clinical trial design has been encourage by both FDA and EMA as a flexible clinical conduct which allows the sponsors to bring successful drugs to the market faster while saving money and resources. As a consequence, for exploratory clinical studies, the conventional, rigid structure for manufacture, warehousing and distribution of Investigational Medicine Products (IMPs) can be advantageously replaced by manufacturing technologies which can tailor the dosage strength with real time adjustment in due course alongside with the open clinical trial and minimize the waste of drug compounds.

In this work, we investigate the potential of industrial inkjet printing technology as a flexible and scalable approach of producing clinical supplies and also overcoming biopharmaceutics challenges of difficult-to-deliver drugs such as poor solubility. A systematic development of poorly soluble model drug was accomplished through (1) rational design for amorphous solid dispersions (ASDs) formulation, (2) inkjet printing process understanding, development and optimization for printing ASD producing ink solution and (3) product design for ASD-printed film that could be filled in hard capsules (PFC).

We will present the results of our ink development approach which selects the desired drug-polymer-solvent systems based on systematic solid-state screening, jetting/printing performance and product key attributes (e.g. dosage size and drug load) and substrate selection. PFC prototypes were produced using a R&D inkjet system and preliminary *in vitro* dissolution and short-term stability studies were carried out. Both the potential and remaining challenges of inkjet printing technology in the context of adaptive drug development will be discussed and summarized.

499034 Tailored Polymers for Structural Application in Large-Scale Extrusion Systems

Abstract:

The explosive growth of inexpensive polymer-based 3D printers across the economy has been driven in part by the promise of true freeform production coupled to digital design. However - as anyone with experience in 3D printing will attest - the hype surrounding the set of technologies known as additive manufacturing often precludes the significant shortcomings and trade-offs inherent in currently available products. In order to realize the full potential of 3d printed products, significant issues in polymer material systems must be addressed.

Material defects are often introduced in 3d printers due to process variability, complex thermal histories due to variable part geometries, sub-optimal toolpaths, process artifacts from machines and software, interfacial chemical reactions between layers and reinforcements, operator errors, and a variety of other variables. Additionally, the set of polymers available for 3D printing is further constrained by practical considerations such as rheology, shelf-life, environmental stability, commercial availability, and cost. These challenges are primarily addressed through either process modification or material system development.

This talk will focus primarily on polymer system development for very large-scale (100 pounds per hour deposition up to about 1000 cubic feet build volume) but will also highlight examples where process and material development are highly dependent on one another. Topics covered will include desired material properties for printing, the role of composite reinforcement materials, composite reinforcement interface and surface chemistry, advanced magnetic multifunctional composite systems, a survey of material systems under development, and highlights of results from recent R&D. Results and highlights will focus on work performed via public-private partnership by the Manufacturing Demonstration Facility (MDF) at Oak Ridge National Laboratory with support from the Advanced Manufacturing Office at the U.S. Department of Energy.

497744 Modeling Flow Phenomena in Fused Filament Fabrication Geometry

Abstract:

The additive manufacturing (AM) field has recently been growing at an exponential rate and is currently transitioning from prototype visualization to functional prototypes and end product development. One of the greatest hurdles to advancing AM is the lack of available materials that can be printed into usable components. Extrusion based additive manufacturing, also known as fused filament fabrication (FFF) and the trademarked name fused deposition modeling (FDM), is the most common method of AM and has the largest catalog of useable materials. However, far fewer options are available when compared to traditional manufacturing methods, which limits the mechanical and chemical properties of parts created by this process. We propose a screening process to efficiently expand the catalog of available materials and reduce the need for extended and expensive testing. The screening process considers the most common failure modes for material extrusion: filament buckling and annular backflow, and integrates a continuum based transport approach to describe the physics of material extrusion through common FFF geometries. Our approach successfully describes the backflow phenomenon, which has not been previously investigated or reported but has been observed as we are developing new materials for the FFF process, including ionomers. We demonstrate the sensitivity of backflow to a materialâ€™s viscosity, the feed rate, and system geometry. Using this approach, we are able to rapidly predict the ability of a polymer to successfully print in an FFF system based on a single experimental measurement and the system geometry. We present the sensitivity of process conditions to the potential for failure to guide process condition selection, including temperature and feed rate.

496776 Influence of Processing on Additively Manufactured Mechanically Adaptive Cellulose Nanocrystal Polymer Composites

Abstract:

Cellulose nanocrystal (CNC) polymer composites have the potential to provide enhanced strength and impart stimuli responsive, mechanically adaptive behavior to a number of commercially available polymeric materials. We are currently investigating the potential to transition these unique characteristics from solvent cast polymer films to functional parts generated using extrusion based additive manufacturing (AM). The relationship between structure, process and properties of functional CNC polymer composites has not been explored when utilizing material extrusion based AM technologies. We are establishing relationships between key material properties such as mechanical strength, thermal transitions and morphology to aid in describing this complex relationship. Developing this understanding of the processing dependent material behavior will aid in developing novel CNC based materials for AM technologies.

In this work, we study a CNC / thermoplastic polyurethane (TPU) nanocomposite. Dynamic mechanical analysis (DMA) experiments were conducted with a submersion chamber attachment, which allows for analysis of mechanical properties in both dry and water saturated environments. Preliminary DMA data indicate processing has a significant impact on dry mechanical properties of the cast films (385 MPa), compression molded films (92 MPa) and extruded fibers (63 MPa). The wet mechanical properties of the cast film, compression molded film and extruded fiber are all very similar (25 MPa ± 5 MPa) indicating that thermal history and orientation has a significant impact on the dry mechanical properties. Geometry also has a substantial impact on rate of diffusion and corresponding rate of mechanical switching, with thin film samples adapting in 350 seconds compared to 15 hours for a 6 mm diameter fiber. Small angle X-ray scattering is used to qualitatively analyze the degree of orientation induced by processing the cast film and fiber samples; despite machine direction orientation in an extruded fiber, we observe reduced dry state storage modulus compared to the unoriented solvent cast film. These results suggest the potential impact of thermal history on surface functionality of the CNCs. We deconvolute the separate, competing contributions of CNC orientation and thermal decomposition; the former is expected to substantially increase modulus, while the latter is expected to reduce coupling between CNCâ€™s.

484122 3D Printed Shape Memory Objects Based on Olefin Ionomer of Zinc-Neutralized Poly(ethylene-co-methacrylic acid)

Abstract:

3D printing enables the net shape manufacturing of objects with minimal material waste and low tooling costs, but the functionality in these objects is generally limited by available materials, especially for extrusion based printing such as Fused Deposition Modeling (FDM). Here we demonstrate shape memory behavior in a complex object, a helix, fabricated by 3D printing with FDM using a commercially available olefin ionomer, Surlyn 9520, which is zinc neutralized poly(ethylene-co-methacrylic acid). The spring-like helix can be mostly straightened to a temporary shape and then recover in 3 dimensions to the initial shape on heating to 70 °C. The performance of 3D printed samples were compared to compression molded analogs using standard shape memory cycles using dynamic mechanical analysis (DMA) with a constant 1 MPa stress for generating the temporary shape. The initial fixity was similar between the printed and compression molded sample, but the recovery was much lower for the 3D printed sample due to the higher strain (lower modulus in the 3D printed part) and higher residual stress. Interestingly, the fixity and recovery on subsequent cycles to the initially recovered permanent shape is greater for the 3D printed than compression molded part. Moreover, the programmed strain can be systematically modulated by inclusion of porosity in the printed part without adversely impacting the fixity or recovery. FDM can be used to directly form thermoplastic shape memory polymers into complex shapes that can be recovered with the appropriate trigger.

494163 Three-Dimensional Printing By Multiphase Silicone/Water Capillary Inks

Abstract:

Three-dimensional (3D) printing of polymers is accomplished easily with thermoplastics as the extruded hot melt solidifies rapidly during the printing process. Printing with liquid polymer precursors is more challenging due to their longer curing times. One curable liquid polymer of specific interest is polydimethylsiloxane (PDMS). We demonstrate a new efficient technique for 3D-printing with PDMS by using a capillary suspension ink containing PDMS in the form of both pre-cured microbeads and uncured liquid precursor, dispersed in water as continuous medium. Owing to the capillary binding of the microbeads, their suspensions bound by small fractions of liquid precursor behaved like pastes, which are flowable at high shear stress and possess high storage moduli and yield stresses that are needed for direct ink writing. The resulting three-phase capillary ink could be 3D printed and cured both in air and under water. The liquid PDMS bridges were thermally crosslinked after printing, resulting in structures that were remarkably elastic and flexible. Their porosity and mechanical properties, such as tensile modulus, could be controlled by the fraction of liquid precursor in the original multiphasic dispersion. Self-standing elastic structures could also be made by directly printing and curing the ink under aqueous solutions. As this ink is made of porous biocompatible silicone, it can be used in 3D printed biomedical products, or in applications such as direct printing of bio-scaffolds on live tissue. The patent-pending method (prov. filed 2016) may also open new opportunities in soft robotics and stimuli-responsive materials.

493399 3D Printing of Polymer-Bonded Magnets Using a Combination of Extrusion Direct Write and Stereolithography Methods

Abstract:

Polymer-bonded magnets have recently gained popularity as a novel magnet manufacturing technique due to its low-cost, design freedom, and magnetizing patterns. The maximum energy product of polymer-bonded magnets is generally lower than that of sintered magnets. However, the fabrication of polymer-bonded magnets does not require high sintering temperature and thus can be more cost effective.

This presentation reports the use of a combination of direct write (DW) and stereolithography 3D printing methods with appropriate process controls to fabricate polymer-bonded magnets at room temperature. Additionally, the presentation will describe and discuss the ink formulation for controlled printing of conformal and 3D magnets. The performance of 3D-printed magnets will be presented and compared to those produced from other manufacturing methods.

490532 PDMS-Based Ink Development for 3D Printing Applications

Abstract:

Additive manufacturing (AM) is an emerging research field due to its ability to manufacture materials with controlled geometry and properties using computer-based 3D printing technology.[1-2] Currently, the main challenge with this technology deals with limited materials, or ink, available for processing, specifically in the field of polymers. Only a single commercially available silicone resin (SE1700) possesses an appropriate viscosity and flow suitable for 3D printing capabilities, severely limiting advancements for improved performance of silicone-based materials for aerospace, engineering, and biomedical applications. Advancements in resin development rely on both new printable formulations and integration of novel fillers in order to alter the materialâ€™s properties. Typical fillers include fumed silica particles; however, other fillers (carbon nanotubes, graphene, surface-modified silica) can be incorporated to target specific applications relying on thermal, mechanical, and/or electronic properties.[3]

In this study, we first report novel 3D printable PDMS-based inks that possess appropriate rheological properties and wide temperature range applicability. Furthermore, compared to commercial silicone kits (e.g. SE1700), these inks also exhibit a very clear composition of ink formulation to broaden our application spectrum. Elastic architectures, which were patterned in a face-centered tetragonal (FCT) structure, were obtained by extrusion from a micro-nozzle using pressurized ink from a syringe, followed by a post-heat treatment to cure the part. The PDMS-based inks were prepared by mixing PDMS with three types of silica fillers having different surface characteristics to achieve three main attributes: i) sufficient elasticity (storage modulus) in the liquid state to withstand typical sagging and flowing issues during 3D printing; ii) excellent processing conditions and engineering properties for 3D printing applications, such as an appropriate pot life (working life or printable time), thermal properties over a wide temperature range (glass transition/crystallization), and a clear ink formulation of the 3D printed products; iii) full understanding of structure-property-performance relationship between silica fillers in the PDMS matrix and their 3D printed products, offering tunable properties to further extend their field of applications. [4]

References

1. S. L. N. Ford, Additive Manufacturing Technology: Potential Implications for U.S. Manufacturing Competitiveness, *J. Int. Commer. Econ.* (2014)
2. *Additive Manufacturing: Opportunities and Constraints: A Summary of a Roundtable Forum Held on 23 May 2013 Hosted by the Royal Academy of Engineering*, Royal Academy of Engineering, (2013)
3. E. B. Duoss, T. H. Weisgraber, K. Hearon, C. Zhu, W. Small, T. R. Metz, J. J. Vericella, H. D. Barth, J. D. Kuntz, R. S. Maxwell, C. M. Spadaccini, T. S. Wilson, *Adv. Funct. Mater.* **24**, 4905 (2014).
4. K.-S. Lee and A. Labouriau *US Provisional Patent Application.* (2017).

500765 3D Printed Active Microfluidic Elements for Portable Bioanalysis Assays

Abstract:

There is an ongoing need for simple, inexpensive, and robust systems that replicate functions of conventional bioanalysis instruments in a format with sufficient portability to deploy in resource-limited settings. A key challenge in the development of these systems, however, involves the ability to dispense precisely metered quantities of liquid reagents and drive their transport through microfluidic networks. Active mechanical pumps offer considerable flexibility compared with passive methods (e.g., capillary wicking in paper-based systems), but sacrifice operational complexity electrical power consumption. Here we describe a novel approach to overcome these limitations by leveraging shape memory polymers to construct a library of fluidic control elements that can be assembled in various arrangements to execute sequential fluidic operations associated with field-based diagnostic assays. These elements (pumps, valves) are easily produced using conventional 3D printing technology, and display a shape memory response within a temperature range that allows them to be operated by exposure to heated water, eliminating the need for battery power and permitting operation by personnel with minimal training.

501190 3D Printed Polymer-Based Bio-Inspired Neural Systems

Abstract:

Polymers are critical materials for the design of biomedical devices and physiological models of biological systems. Additive manufacturing now provides the ability to process biocompatible polymers into complex geometries for a variety of biological applications. However, the design of functional and higher order scaffolds and platforms for regeneration and modeling of the nervous system requires the ability to interweave multiple polymers into bio-inspired architectures. Here, we address the manufacturing of multi-functional anatomical scaffolds for peripheral nerve regeneration and *in vitro* platforms for modeling higher order pathophysiological responses in the nervous system via a novel multi-material microextrusion-based 3D printing process. Printing of polymers with disparate material properties is discussed with a focus on critical polymer properties, such as polymer rheological properties, melting temperature, and gelation temperature. The integration of multiple polymers into bio-inspired architectures, including elastomers, thermoplastics, and hydrogels, will be discussed. The manufacturing process is also compatible with the 3D printing of neuronal and non-neuronal cells at high viability. 3D structured-light scanning is used to construct anatomical models of peripheral nerve that inform the programming of path information for scaffold printing. Ultimately, polymer-based additive manufacturing appears poised as a unique processing approach for the design and manufacturing of novel multi-functional microphysiological neural systems.

AUTHOR INDEX

AUTHOR INDEX

AUTHOR INDEX

AUTHOR INDEX

AUTHOR INDEX

AIChE
120 Wall Street, FL 23
New York, NY 10005-4020

ISBN 978-1-5108-5784-1
9 781510 857841

CURRAN ASSOCIATES INC.
proceedings.com

17th Topical Conference on Gas Utilization 2017

Topical Conference at the 2017 AIChE Spring Meeting and
13th Global Congress on Process Safety

San Antonio, Texas, USA
26 – 30 March 2017

AIChE
The Global Home of Chemical Engineers